Exercise 4[1.19.1] [on solutions of $y' + p(x)y = r(x)$]

Let p and r be continuous functions on I. Let $x_0 \in I$, $y_0 \in R$. The solution function satisfying the general first-order linear equation $y' + p(x)y = r(x)$ and the initial condition $y(x_0) = y_0$ may be specified as

p. 58

$$y = \exp\left[-\int_{x_0}^{x} p(u)\, du\right]\left[\int_{x_0}^{x} r(u) \exp\left[\int_{x_0}^{u} p(v)\, dv\right] du + y_0\right].$$

Theorem 2.1.1 [on solutions of $y'' + p(x)y' + q(x)y = r(x)$]

Let p, q, r be continuous functions on I.
Let $x_0 \in I$; $y_0, y_0' \in R$.

Then there is a unique function on I such that

p. 61

$y(x_0) = y_0$,

$y'(x_0) = y_0'$,

and $y(x)$ satisfies the equation

$$y'' + p(x)y' + q(x)y = r(x) \qquad \text{for} \quad x \in I.$$

Theorem 2.5.1 [on solutions of $L(y) = y'' + p(x)y' + q(x)y = 0$]

Let f_1 and f_2 be linearly independent solutions of $L(y) = 0$ on the interval I.
Let f_3 be any other solution of $L(y) = 0$ on I.

Then

p. 71

$$f_3(x) = c_1 f_1(x) + c_2 f_2(x) \qquad \text{for some} \quad c_i \in R.$$

The general solution is

$$y(x) = c_1 f_1(x) + c_2 f_2(x).$$

DIFFERENTIAL EQUATIONS

THEORY AND USE IN TIME AND MOTION

DIFFERENTIAL EQUATIONS

THEORY AND USE IN TIME AND MOTION

ALICE (B. DICKINSON

(raunlich)

Smith College

ADDISON-WESLEY PUBLISHING COMPANY

Reading, Massachusetts · Menlo Park, California · London · Don Mills, Ontario

This book is in the

ADDISON-WESLEY SERIES IN MATHEMATICS

Consulting Editor
Lynn H. Loomis

Copyright © 1972 by Addison-Wesley Publishing Company, Inc. Philippines copyright 1972 by Addison-Wesley Publishing Company, Inc.

All rights reserved. No part of this publication may be reproduced, stored in a retrieval system, or transmitted, in any form or by any means, electronic, mechanical, photocopying, recording, or otherwise, without the prior written permission of the publisher. Printed in the United States of America. Published simultaneously in Canada. Library of Congress Catalog Card No. 77-136120.

PREFACE

My intent is to offer a student with one year of calculus an introduction to the origins, theory, and implications of differential equations, and to give particular attention to the transition from the physical situation to the mathematical model, and vice versa. Mathematicians and scientists often assume that each other consider these aspects, while, in fact it is seldom that either do. The structure is provided by two lines of development. Differential equations of increasing complexity serve as models for different aspects of two basic problems, (a) the understanding of motion, and (b) the measurement of time. The necessary physical concepts are discussed. (In my own course, films and occasional physical demonstrations by students or faculty supplement these discussions.) The theory is intended to reinforce the calculus and to anticipate some advanced mathematics, so that differential equations are seen as a part of the mainstream rather than operational offshoots. I believe that mathematics is learned by doing, and that the excitement comes from discovering, rather than accepting, principles. Hence some of the development is in the exercises; the amount increases in each successive chapter. In a one-semester course I have considered Chapters 1–3 in class. As a final project, students interested in theory have worked through the existence theorems in Chapter 4 independently. Others have preferred to consider applications in their own fields. (One of my criteria for such individual papers is that they be lucid enough for me to understand.)

It is hoped that this differential equations book may serve as an introduction to applied mathematics and at the same time as an introduction to the theorem–proof style of mathematics.

My own commitment to mathematics stems not from the undergraduate classroom, but from the M.I.T. Radiation Laboratory and the Sperry Gyroscope Research Laboratories during World War II. It was there that I first saw differential equations as models of physical situations and knew the excitement of understanding some of my own experience and knowledge as realizations of these models. I believe that making the model idea explicit is useful in undergraduate mathematics. In the process of making assumptions and constructing equations, one sees mathematics as a language of relations and structures that provides insights in a variety of fields.

Differential equations is not my field, but I have sometimes been asked to teach an introductory course in this subject. Student response to my attempts to blend theory and applications suggested, and indeed constructed, the book. In particular, Ellen Borie and Joan Hutchinson made thoughtful critiques of the early version; Marie Chow, Susan Mellen, and Marlyn Reynolds offered constructive suggestions in the later drafts.

I am grateful for the understanding conveyed by E. Kamke's *Differential-gleichungen Reeler Funktionen*; for detailed conversations with David Park, professor of physics at Williams College, which converted an idea into a plan of action; for the insights into teaching which Emil Artin, Carl Garabedian, T. H. Hildebrandt, G. Y. Rainich, Earl Rainville, and R. L. Wilder conveyed by example and which I have pondered during years of teaching and during the writing of this book; for the facilities of the Cambridge University Libraries; for the assistance of Mr. E. E. Makepeace of the Manchester Public Libraries; and for a family who, male and female, young and old, shared the daily work so that each of us could cultivate our own projects.

Ashfield, Massachusetts   A.B.D.
January 1972

CONTENTS

Chapter 3   Series Solutions

Chapter 4   Existence Theorems

General Reading Suggestions

Appendix A

Appendix B

Appendix C

Appendix D

CHAPTER 1

FIRST-ORDER EQUATIONS

1.1 INTRODUCTION

A differential equation states a condition on the derivatives of a function, or functions. A solution to a differential equation is a function, or set of functions, which satisfies the condition. For example, $f' = f$ states that a function must equal its derivative function. The function $f(x) = e^x$ on R, the set of real numbers, is a solution. Functions are the basic elements.

Functions are specified in various ways. The function f, which associates with each x in R the number x^2, may be specified by

1) the relation $f(x) = x^2$, $x \in R$;
2) the mapping $f: x \rightarrow x^2$, $x \in R$;
3) the name, the square function on R;
4) the set of ordered number pairs $\{(x, f(x)): f(x) = x^2, x \in R\}$;
5) the graph $\{(x, y): y = x^2, x \in R\}$. (This last set is the same as that in (4); it differs only in its traditional interpretation as a set of points.)

Each notation emphasizes a particular aspect of the function concept and is useful in certain situations. All are used widely.

The domain is essential in specifying a function. For example, $f_1: x \rightarrow x^2$, $x \in R$ and $x \geq 0$, is an increasing function; $f_2: x \rightarrow x^2$, $x \in R$, is not. The function $f_3(x) = 1/x$, $x \in R$ and $x \geq 1$, is a bounded function (see Appendix A.1); the reciprocal function on R^+, the positive reals, is not. For a function f with domain I, the range, or set $\{f(x): x \in I\}$, is denoted as $f[I]$.

A solution function must be differentiable over its domain. Otherwise it could not satisfy a differential equation. Moreover, solution functions are restricted to functions whose domain is precisely an interval. This restriction derives from the source of differential equations, that is, questions concerning continuous motion. One of the dominant questions in Newton's time (seventeenth century) concerned the motion of the planets about the sun. The very desire to explain the observed motion by a fundamental law, and to predict future motion, reflects the intuitive assumption of continuous motion in an unbroken time interval. The mathematical model of such motion is a continuous function on an interval. The numbers in the domain and range represent time intervals and distances, respectively. Functions defined at isolated values or on disconnected intervals are of no

1

value in predicting continuous motion. Differential equations still provide models of continuous phenomena in physics, chemistry, biology, and the social sciences.* Hence the relevant mathematical functions must be defined over a connected piece of the real line, with no breaks or missing numbers. That is, the domain must be a single real interval I.

The transition from a physical situation to a mathematical model, and the physical implications of the differential equations and their solutions, occupy a significant part of the development. The physical situations considered here are centered about two basic problems, the study of motion and the measurement of time.

For an object in motion, the velocity and/or acceleration are often known. Since velocity and acceleration are defined as derivatives with respect to time, these conditions on the motion can be written as a differential equation. For an equation stating conditions on $v(t) = \dot{s}(t)$ and $a(t) = \ddot{s}(t)$ (dots over letters indicate derivatives with respect to time), the desired solution is a function $s: t \rightarrow s(t)$ on I which describes the motion by specifying a position s for any time $t \in I$. Before setting up a differential equation as a model for a particular physical situation, a mathematical structure sufficient to provide solutions and to establish a notation will be developed.

Just as algebraic equations depend on relations and operations defined for numbers, differential equations require that the usual relations and algebraic operations be defined for functions. *Two functions f_1, f_2 are equal if they have the same domain B and $f_1(x) = f_2(x)$ for each $x \in B$.* *The sum of two functions* on a common domain B is defined as $f_1 + f_2: x \rightarrow f_1(x) + f_2(x)$, $x \in B$. Similarly, the *product function* is $f_1 f_2: x \rightarrow f_1(x)f_2(x)$, $x \in B$. The definition of equality implies that the differential equation $f' = f$ is equivalent to the numerical equation $f'(x) = f(x)$ for all x in the domain of f. Note that if f is a differentiable function, f' is defined on the same domain.

Functions have properties similar to those of the real numbers. For functions on a domain B, the additive identity is the zero function $0_f: x \rightarrow 0$, $x \in B$. The multiplicative identity is the unit function $1_f: x \rightarrow 1$, $x \in B$. For each real number r, there is a constant function $r_f: x \rightarrow r$, $x \in B$.

Quite apart from algebraic properties, some functions have an inverse. A function f, defined by the set of number pairs $\{(x, (f(x)), x \in B\}$, has an *inverse function* $f^{\leftarrow}: f(x) \rightarrow x$, $f(x) \in f[B]$ if the set

$$\{(f(x), x): f(x) \in f[B]\}$$

defines a function. If $y = f(x)$, the inverse function is indicated both by $f^{\leftarrow}$ and $y^{\leftarrow}(y) = x$. Note that $(f^{\leftarrow})^{\leftarrow} = f$.

---

*A theory of discontinuities to model such instances as the breaking of a wave or the division of a cell is just now being developed by René Thom in Paris and Christopher Zeeman in England. The mathematics is quite sophisticated. Thom's exposition of this theory is scheduled to be published by Benjamin.

EXERCISES [1.1.1]

· ·

1. Define the additive inverse function $-f$ of f.
2. Define the multiplicative inverse function $1/f$ of f.
3. Are the domains of $-f$ and $1/f$ always the same as that of f?
4. Sketch the functions below, together with their additive inverse, multiplicative inverse, and inverse functions. If necessary, restrict the domain to obtain an inverse or multiplicative inverse function.

   a) $f_1(x) = 2$, $x \in R$ b) $f_2(x) = x$, $x \in R$ c) $f_3(x) = \sin x$, $x \in R$
   d) $f_4(x) = e^{-x}$, $x \in R$ e) $f_5(x) = 1/x$, $x \in R^+$ f) $f_6(x) = \ln x$, $x \in R^+$
   g) $f_7(x) = e^{-|x|}$, $x \in R$

5. Sketch the sums and products of

   a) f_1 and f_2,   b) f_1 and f_3,
   c) f_2 and $-f_3$,   d) f_3 and f_4,
   e) f_3 and $f_5^{\leftarrow}$,   f) f_3 and f_7,

   where each f_i is as defined in Exercise 4. If necessary, modify domains so that products and sums can be defined.

· ·

It would be convenient and intuitively satisfying to write differential equations as functional equations, such as $f' = f$, especially since much of the discussion and point of view herein concerns functions. But alas, differential equations enjoyed two centuries of development before reforms in notation and definition led to the present concept of a function. It would be as easy to change the spelling rules of the English language. Earlier mathematicians depended largely on intuition; to the outsider they were magicians. Mathematics was not for public consumption and precise notation seemed unimportant. The tradition of writing differential equations as numerical equations became well established. And today these equations are among the cast-iron idioms* of mathematical language. The functional notation could be achieved by adopting the symbol i for the identity function $i: x \rightarrow x$, $x \in R$. Then, for example, the square and exponential functions would be denoted as i^2 and e^i. But changes in notation inhibit easy browsing in the vast and rich literature of differential equations that now exists. Hence numerical differential equations continue to be used.

The usual convention in notation is that if f indicates the function $f: x \rightarrow y$, then $f(x) = y(x) = y$ denotes the number associated with x by the function f. Thus, the numerical equations

$$f'(x) = f(x), \qquad y'(x) = y(x), \qquad y' = y,$$

---

*In Fowler's Modern English Usage, idiom is characterized as "conservative, standing in the ancient ways, . . . , permitting no jot or tittle of alteration in the shape of its phrases."

for all x in the domain of f, state the same conditions and have the same solution functions. Equations or systems of equations, algebraic or differential, are *equivalent* if they have the same set of solutions. Each of the three equations above is equivalent to the functional equation $f' = f$. The functional equation $f' = i$ is equivalent to the numerical equations

$$f'(x) = x, \qquad y'(x) = x, \qquad y' = x,$$

for all x in the domain of f. The briefest form, $y' = x$, is the one most commonly used.   Other derivative notations are $y'(x) = dy/dx = D_x y$.   Hence the same equation may be written as $dy/dx = x$ or $D_x y = x$.

EXERCISES [1.1.2]

. . . .

Write the numerical differential equations as functional equations, and the functional equations as numerical equations.

1. $y''^2 + y'^2 = 1$

2. $f'^2 + 1_f = 0_f$

3. $y'' + 2y' + 3y = 0$

4. $if' = e^i$

5. $2\dfrac{dy}{dx} + y = 0$

6. $f' = i^2 + 2_f i$

7. $x^2 \dfrac{d^2 y}{dx^2} + 2x\dfrac{dy}{dx} - 3y = x^3$

. . . .

Hereafter differential equations will be written as numerical equations.   But remember that the unwritten phrase "for every x in the domain of the function" makes these numerical equations equivalent to functional equations.

Once a differential equation is set up as a model of a physical situation, the mathematical problem is to determine solutions or information concerning solutions.   Consider solutions of specific equations.

1) $y' = y$ is satisfied by $f_1(x) = e^x$, $x \in R$.   The functions $f_2: x \to e^{2+x}$ and $f_3: x \to 3e^x$, both on R, are also solutions.   Are there more solutions?   How many?

2) $y''^2 + y'^2 = 1$ is satisfied by both the sine and cosine functions on R.   Are there other solutions?   Is is possible to obtain the set of all solutions?

3) $y'^2 + 1 = 0$ has no solutions among the real functions.

These simple differential equations raise basic questions.   Does a given equation have a solution?   Are there conditions that ensure the existence of a solution?   How many solutions are there?   Is it possible to obtain all the solutions?   Is each solution a special case of a general solution function? Is there a standard form in which all the solutions may be written? Under what conditions is there a unique solution?

The answers to these questions make up the *existence* and *uniqueness theorems* which provide an orderly development of the subject. The question of unique solutions is of particular interest in applications. A scientific experiment is acceptable only if it can be repeated; that is, the same conditions must lead to the same results. The mathematical model of such a situation should lead to a unique solution function representing the expected results.

How are solutions determined? There are large collections of ingenious methods for solving specific differential equations. These are important, indeed. But such specialized techniques can be developed by the individual when the need or interest arises. Here, only the basic theory and general methods for large classes of equations are considered.

1.2 FIRST-ORDER EQUATIONS AND DIRECTION FIELDS

The *order* of a differential equation is the number indicating the highest-order derivative in the equation. The first-order equation $y' = 1$ can be solved at sight. A solution is any function such as $y_1(x) = x + 2$, $x \in R$, whose derivative is the constant function 1. What are some solutions of $y' = x$? This question, in slightly different form, is familiar to calculus students. What are the antiderivatives of the identity function $f(x) = x$?

The conditions stated by these first-order equations have a geometric interpretation in terms of slopes. For example, the equation $y' = x$ indicates that the solution curve has slope $y'(x) = x$ at each of its points (x, y). Sketch segments with slope x at enough points (x, y) so that a pattern emerges. Figure 1.1 represents the *direction field* of the equation $y' = x$. Note that a direction field indicates *slopes* rather than directions, but again the terminology is well established. A similar

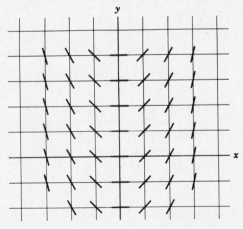

Figure 1.1

notion in physics is the force field which indicates the direction and magnitude of a force acting on a unit object at particular points in a region.

A smooth curve that fits this direction field on an interval I (Fig. 1.2) is the

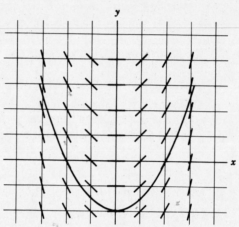

Figure 1.2

curve of a differentiable function on I satisfying the equation $y' = x$. The curve and the function are specified by the same set of number pairs; the terms "solution function" and "solution curve" are used interchangeably.

The direction field in Fig. 1.1 suggests that parabolas, symmetric to the vertical axis, are solution curves of $y' = x$. These curves correspond to the solutions defined by the set of antiderivatives $\{y_c(x) = \frac{1}{2}x^2 + c, x \in I, c \in R\}$. The solution curve through $(0, -2)$ represents the function $y_2(x) = \frac{1}{2}x^2 - 2$. Are there other types of solution curves? It seems clear that certain curves, such as lines, do not fit this particular direction field. However, it is difficult, using only geometric intuition, to eliminate all other possibilities. The question may be answered by turning to the basic theorems of calculus.

EXERCISES [1.2.1]

. . . .

Sketch direction fields and approximate solution curves for the following equations. In each case, indicate graphically and, if possible, analytically particular solution curves through each of the given points.

1. $y' = 1$; $(1, 0)$, $(-2, 3)$

2. $y' = y$; $(0, 1)$, $(1, e)$, $(1, 0)$

3. $y' = y^{1/2}$; $(0,1)$, $(1, 0)$, $(1, -1)$

4. $y' = x^2$; $(0, 1)$, $(1, 0)$

5. $y' = 1/x$; $(1, 0)$, $(0, 1)$, $(-1, 1)$

6. $y' = y/x$; $(1, 0)$, $(0, 1)$, $(1, 1)$

7. $y' = e^x/x$; $(1, 0)$, $(-1, 1)$

. . . .

1.3 THE EQUATION $y' = f(x)$

The first-order equations $y' = 1$ and $y' = x$ are special cases of the general equation $y' = f(x)$, where f is any specified function on an interval I. The direction field for this general equation is well defined for all points (x, y), where $x \in I$. Does every such equation have a solution? Or, rephrased, does every function f have an antiderivative? Intuitively, if the direction field is continuous, solution curves may be visualized. But if the direction field is not continuous, that is, if the slope changes erratically rather than gradually as x varies, a differentiable (or smooth) curve will not fit.

The key to the solution of the general equation $y' = f(x)$ is the integral function $A(x) = \int_a^x f(u)\, du$ on I and the theorems relating integrals, derivatives, and antiderivatives. These theorems from the calculus will be stated and discussed briefly. . The reader is advised to review the relevant material. The review of theorems as used leads to the assimilation of a useful body of mathematics, as well as an understanding of the overall structure.

Let f be a continuous function on an open interval I in R. Then the following statements hold:

S(1.3.1) The number $\int_a^b f(x)\, dx$, called the Riemann integral, is well defined for $a, b \in I$. (See Appendix A.4.)

S(1.3.2) The function $A(x) = \int_a^x f(u)\, du, a, x \in I$, is differentiable and $A'(x) = f(x)$.

S(1.3.3) If G and H are differentiable functions on I such that $G'(x) = H'(x)$ on I, then $H(x)$ and $G(x)$ differ by a constant on I.

S(1.3.4) If F is any antiderivative of the function f on I, then

$$\int_a^b f(x)\, dx = F(b) - F(a), \qquad \text{for} \quad a, b \in I.$$

Statement S(1.3.1) is a basic theorem of integral calculus. Although the machinery of partitions and Riemann sums (or upper and lower sums) is set up in elementary calculus, the actual proof of the theorem is usually deferred to advanced calculus or real-variable theory. Here, S(1.3.1) will be taken as an assumption on which to build. The interpretation of $\int_a^b f(x)\, dx$ as the algebraic area between the x-axis and the curve of f on the interval $[a, b]$ will be useful. The continuity of f is stronger than necessary for S(1.3.1), but weaker conditions would not suffice for S(1.3.2) and S(1.3.3). (See Exercise 10 [1.3.1].)

Statement S(1.3.2) exhibits integration and differentiation as inverse operations on a function. This fundamental theorem and its proof should be part of one's indispensable mathematical baggage. Statement S(1.3.3) is a direct consequence of the mean-value theorem. Statement S(1.3.4), implied by S(1.3.2) and S(1.3.3), offers a method of evaluating definite integrals in terms of antiderivatives, rather than as limits of Riemann sums.

What are the implications of these theorems for the differential equation $y' = f(x)$, if f is continuous on the open interval I? Statement S(1.3.2) implies that $y(x) = \int_a^x f(u)\, du$, a, $x \in I$ [well defined by S(1.3.1)] is a solution. Statement S(1.3.3) implies that any other solution may be specified as

$$y_c(x) = \int_a^x f(u)\, du + c, \qquad a, x \in I.$$

And the basic rules of differentiation imply that every such function is a solution. Hence the set of all solutions is precisely the set of integral functions

$$\left\{ y_c(x) = \int_a^x f(u)\, du + c : a, x \in I, c \in R \right\}.$$

Solutions are often called the *integral curves* of a differential equation.

Statement S(1.3.4) provides a method of expressing the integral solution in terms of well known elementary functions, if an antiderivative is known. For many functions, an antiderivative in terms of elementary functions is not known. But the integral function is well defined [S(1.3.1)], and tables of approximate function values may be obtained by numerical methods. Such numerical approximations have long been used in applications. And, today, numerical methods are significant research techniques in differential equations.

Most of the questions concerning solutions of $y' = f(x)$ have been answered. But there are as many solutions as there are real numbers. Is there a useful way to classify these solutions? What sort of added restriction would sort out a particular solution? The direction field for $y' = x$ (Fig. 1.1) suggests that a single solution curve passes through each point. Does this situation obtain in general? Statement S(1.3.3) implies that at most one integral curve passes through a point in the region $\{(x, y): x \in I, y \in R\}$. For two integral curves through (x_0, y_0) differ by a constant, and at (x_0, y_0) the constant difference is zero. Moreover, each (x_0, y_0) is on the solution curve

$$y(x) = \int_{x_0}^x f(u)\, du + y_0, \qquad x_0, x \in I.$$

Hence there is precisely one integral curve through each point in the region $\{(x, y): x \in I, y \in R\}$. Rephrased, there is a unique solution satisfying the differential equation and the condition $y(x_0) = y_0$.

Together, these results provide a theorem which, as well as dealing with existence and uniqueness, offers a simple form in which to write solutions.

Theorem 1.3.1 [on solutions of $y' = f(x)$]

Let f be a continuous function on the open interval I.
Let (x_0, y_0) be any point such that $x_0 \in I$, $y_0 \in R$.

Then there exists a unique function on I satisfying the differential equation $y' = f(x)$ and the condition $y(x_0) = y_0$.

This solution may be written in the form

$$y(x) = \int_{x_0}^{x} f(u) \, du + y_0, \qquad x \in I.$$

This theorem, applied to the specific equation $y' = x$, states that through any point (x_0, y_0) in the plane, there passes a unique solution curve

$$y(x) = \int_{x_0}^{x} u \, du + y_0, \qquad x \in R.$$

Since, in this case, an elementary antiderivative function is known, the solution may be expressed as

$$y(x) = \tfrac{1}{2}x^2 - \tfrac{1}{2}x_0^2 + y_0, \qquad x \in R.$$

EXERCISES [1.3.1]

. . . .

1. Compare the two sets of solutions obtained for the equation $y' = x$: the set of anti-derivative functions

$$\{y(x) = \tfrac{1}{2}x^2 + c, x \in R : c \in R\}$$

corresponding to the curves on the direction field in Fig. 1.2, and the solution curves

$$y(x) = \tfrac{1}{2}x^2 - \tfrac{1}{2}x_0^2 + y_0, \qquad x \in R,$$

on each (x_0, y_0) of the plane. Is there a 1–1 correspondence between the sets? What is the relation between the constants x_0, y_0, and c?

For the equations in Exercises 2 through 9, specify:

    a) The continuous function f on a largest possible interval I.
    b) A solution satisfying the condition $y(x_0) = y_0$ where $x_0 \in I$.
    c) A particular solution through each of the given points.
    d) Compare the solutions of the equations in Exercises 3 through 6 with the curves on the corresponding direction fields in Exercises [1.2.1].

2. $y' = x$; $(0, -2)$, $(2, 0)$ 3. $y' = 1$; $(1, 0)$, $(-2, 3)$

4. $y' = x^2$; $(0, 1)$, $(1, 0)$ 5. $y' = 1/x$; $(1, 0)$, $(0, 1)$, $(-1, 1)$

6. $y' = e^x/x$; $(1, 0)$, $(-1, 1)$ 7. $y' = \cos x$; $(0, 1)$, $(\pi/2, 1)$

8. $y' = \tan x$; $(0, 0)$, $(0, 1)$, $(\pi/2, 1)$, $(2\pi/3, 0)$

9. $y' = |\sin x|$; $(0, 0)$, $(3\pi/2, 0)$

10. Show that statements S(1.3.2) and S(1.3.3) may not hold for discontinuous functions.

    a) Sketch the step function

$$\begin{aligned} f(x) &= 0 \text{ on } [0, 1) \\ &= \tfrac{1}{2} \text{ on } [1, 2) \\ &= 1 \text{ on } [2, 3]. \end{aligned}$$

b) Sketch the function $A(x) = \int_0^x f(u)\, du$ on $[0, 3]$. Let

$$\int_0^n f(u)\, du = \lim_{x \to n} \int_0^x f(u)\, du$$

for $n = 1, 2, 3$. Is the function defined by $A(x)$ differentiable on $[0, 3]$?

c) Define functions G and H on $[0, 2]$ such that
   i) G and H are differentiable except at $x = 1$;
   ii) $G'(x) = H'(x)$ except at $x = 1$;
   iii) $G(x) - H(x) \neq C$, a constant, on $[0, 1)$ and $(1, 2]$.

d) Does the conclusion of Statement S(1.3.3) hold if G and H are differentiable and $G'(x) = H'(x)$ on a domain B (not necessarily an interval) such as $B = (0, 1) \cup (2, 3)$?

. . . .

1.4 NUMERICAL METHODS

Most of the solutions above may be written in terms of elementary functions. But the integral solution of the equation in Exercise 6 can be further specified only by a table of numerical values. Before computing values for $y' = e^x/x$, numerical methods will be applied to the familiar equation $y' = x$ with condition $y(0) = -2$, so that the numerical approximations may be compared to the solution already obtained.

The simplest procedure, known as *Euler's method*, uses the slope at a known point (x, y) to estimate the function value $y(x + h)$. As a limiting value, the first derivative is approximated by the difference quotient, that is, for h small:

$$\frac{y(x + h) - y(x)}{h} \doteq y'(x) \quad \text{or} \quad y(x + h) \doteq y(x) + hy'(x).$$

The smaller the h, the better the approximation. The differential equation $y' = x$ and condition $y(0) = -2$ specify a point $(0, -2)$ on the solution curve and a slope $y' = 0$ at that point. Thus $y(0.2)$ may be estimated as

$$y(0 + 0.2) = y(0) + 0.2y'(0) = -2.$$

The procedure is repeated at each new point; h is usually held constant. The computations and sequence of function values on $[0,2]$ for $h = 0.2$ are shown in Table 1.1.

Since $y'(x)$ changes throughout each interval $(x, x + h)$, a better approximation is obtained by using the average of the derivatives at the endpoints, or the derivative at the midpoint. (See Exercises 3 and 4 below.)

TABLE 1.1

x	$x + h$	$hy'(x) = 0.2x$	$y(x)$	$y(x + h)$
0			-2	
0	0.2	$0.2(0) = 0$	-2	-2
0.2	0.4	$0.2(0.2) = 0.04$	-2	-1.96
0.4	0.6	0.08	-1.96	-1.88
	0.8	0.12		-1.76
	1.0	0.16		-1.60
	1.2	0.20		-1.40
	1.4	0.24		-1.16
	1.6	0.28		-0.88
	1.8	0.32		-0.56
	2.0	0.36		-0.20

EXERCISES [1.4.1]

1. a) Plot the numerical approximation in Table 1.1 on the direction field of the equation $y' = x$.
   b) Compare the numerical approximation with the solution function through $(0, -2)$. What are the errors at $x = 1$ and $x = 2$?

2. a) Obtain a numerical approximation for $y' = x$ and $y(0) = -2$ on $[0, 2]$ with $h = 0.1$.
   b) Compare (a) with the solution function and the numerical approximation with $h = 0.2$ (Table 1.1).

3. Show that for $y' = x$ and $y(0) = -2$, the two modifications of Euler's method, one using the average of slopes at the endpoints of each interval and the other using slopes at the midpoint, are equivalent.

4. Use one of the modifications in Exercise 3 to obtain a numerical approximation for $y' = x$ and $y(0) = -2$ on $[0, 2]$ with $h = 0.2$. Compare the accuracy with that of the other numerical approximations. [The accuracy, in this case, is due to the form of the solution function. Show that for $y(x) = \frac{1}{2}x^2 - 2$, $y(x + h) = y(x) + hy'(x + \frac{1}{2}h)$.]

5. Consider the equation $y' = e^x/x$ and condition $y(1) = 0$.

   a) Does a solution curve exist? What is the largest domain of the solution? Is the solution function differentiable? If so, what is the derivative? Is the solution increasing or decreasing? Is the solution curve concave upward or downward?
   b) Obtain an approximation on $[1, 2]$. Use exponential tables and two-place accuracy. Let $h = 0.1$.
   c) Does the approximation fit the direction field?
   d) An easy method of checking computations is to list the first differences, $y(x + h) - y(x)$, as each new function value is obtained. Any erratic changes in the

first differences usually indicate a mistake in computation. Check the first differences on some of the numerical solutions above.

*6. a) Use Simpson's (1710–1761) rule to obtain numerical approximations for the integral solutions of equations $y' = x$ and $y' = e^x/x$ with the given conditions $y_0 = y(x_0)$. Use intervals $[0, 2]$ and $[1, 2]$, respectively, and $h = 0.1$.

   b) Estimate the error in $y(2)$ for both approximations in (a). A bound for the error on an interval $[a, b]$ is

$$E \le \frac{(b + a)}{180} kh^4,$$

where $k \ge f^{(4)}(x)$ for all x in $[a, b]$. The superscript (4) indicates the fourth derivative.

*7. A computer may take 10^{-10} sec to perform an addition and 10^{-9} sec for a multiplication. Estimate, for both equations, $y' = x$ and $y' = e^x/x$, how many values of $y(x)$ may be computed in one minute, using the Euler method, the modified Euler method, and Simpson's rule.

· ·

The solution for $y' = e^x/x$ and a particular condition $y(x_0) = y_0$ cannot be obtained explicitly in terms of elementary functions, but numerical values may be tabulated. Note that an approximation varies with the choice of step size h, the particular numerical method, and even the number of digits retained. Thus there may be many numerical approximations with differing degrees of accuracy. More numerical methods will be introduced as the necessary techniques are developed. Questions of accuracy will be discussed in Chapters 3 and 4.

Many useful functions have originated as integral solutions of differential equations. The study of their properties is the area of mathematics known as *special functions*. Some of these special functions will be discussed in particular applications. Extensive numerical tables have been prepared for many special functions. Some are published, others are indexed and catalogued and are available on request.

1.5 FALLING BODIES

The simplest motion is one-dimensional, as exemplified by a falling body. Falling objects elicit physical reactions, such as catching or moving out of the way, that reflect personal predictions of the path of motion. In order to obtain a quantitative description of such motion, let $t = 0$ indicate the beginning of the fall and $t = t_c$ the end, or time of collision. Let h, v, and a indicate the height, velocity, and

---

*The symbol * indicates exercises, sections, or discussions which are difficult or theoretical and may be omitted without loss of continuity.

acceleration at any time t during the fall. If the functions defined by $h(t)$ and $v(t)$ on $[0, t_c]$ are differentiable, then by definition $\dot{h}(t) = v(t)$ and $\dot{v}(t) = a(t)$.

One simple model is a differential equation based on the observation of Galileo (1564–1642) that objects fall with constant acceleration. If height above the earth is represented by a positive number, then acceleration toward the earth is represented by a negative number. Thus Galileo's law, $a = g$, stated as a differential equation, is $\dot{v} = g$, where g is a negative constant. The initial condition (specifying a function value on a time domain) $v(0) = v_0 = 0$ indicates that the object is released rather than projected.

Given the equation $\dot{v} = g$ and initial condition $v_0 = 0$, does Theorem 1.3.1 apply? The hypothesis asks that f be continuous on an open interval. $\dot{v}(t) = f(t) = g$ is a constant function and hence continuous on $(0, t_c)$, but the initial condition concerns the endpoint 0, and $\dot{v}$ is not continuous on any open interval containing 0 (see Fig. 1.3). Can the theorem be extended to a closed interval?

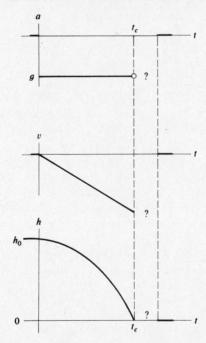

Figure 1.3

The motion just after collision is complicated; it depends on the elasticity of the materials. However, if the model is restricted to $[0, t_c]$ and no predictions are made for $t > t_c$, it is not unreasonable to assume that $a(t_c) = g$. Then $\dot{v}(t) = f(t) = g$ is continuous on $[0, t_c]$. (See Appendix A.2.) But is the expected solution, $v(t) = \int_0^t g \, du$, differentiable at the endpoints? Or, in general, if the first sentence in Theorem 1.3.1 is changed to read "Let f be a continuous function on the closed interval I," does the theorem still hold? Theorem 1.3.1 depends on Statements

S(1.3.1), S(1.3.2) and S(1.3.3). Are these statements valid on a closed interval? S(1.3.1) certainly holds since it concerns only a closed interval $[a, b]$ in I. The difficulty in S(1.3.2) is the existence of derivatives at the endpoints of a closed interval. If the notion of derivative is extended to include right-handed or left-handed derivatives at the endpoints, then the integral function is differentiable on the closed interval I. (See Appendix A.3.) S(1.3.3) still obtains because the mean-value theorem concerns a continuous function on a closed interval and requires differentiability only on the open interval. Consequently Theorem 1.3.1 is valid on a closed interval if differentiation is extended to the endpoints.

Open intervals are natural domains for derivative functions since two-sided limits are involved. Integral functions are often defined for all values between and including two fixed points. A compromise is needed when both derivative and integral functions are involved. Another procedure is to obtain solutions on the open interval and then define function values at the endpoints by the limit convention. For example, let

$$v(0) = \lim_{t \to {}^+0} v(t) \quad \text{and} \quad v(t_c) = \lim_{t \to {}^-t_c} v(t).$$

In effect, the initial condition is imposed as a limiting condition to single out one solution among a set of solutions.

Theorem 1.3.1, extended to a closed interval, implies that

$$v(t) = \int_0^t g \, du + 0 = gt, \qquad t \in [0, t_c]$$

is the unique solution function satisfying the differential equation $\dot{v} = g$ and the initial condition $v(0) = 0$. The physical implication is that during the fall, the velocity is a constant multiple of the time elapsed.

This solution gives rise to the differential equation $v(t) = \dot{h}(t) = gt$. The height at release imposes the initial condition $h(0) = h_0$. The linear function $f(t) = gt$ is continuous on $[0, t_c]$, and Theorem 1.3.1 applies again. The unique solution is

$$h(t) = \int_0^t gu \, du + h_0 = \tfrac{1}{2}gt^2 + h_0, \qquad t \in [0, t_c].$$

The description of the motion is now complete. The position, velocity, and acceleration values are determined at every instant of the fall. In particular, for $v_0 = 0$,

$$h(t) = \tfrac{1}{2}gt^2 + h_0,$$
$$v(t) = gt,$$
$$a(t) = g.$$

Another physical implication of these solutions concerns the *particle velocity*. The velocity of the object at any height $h \in [0, h_0]$ may be expressed in terms of the distance fallen, $(h_0 - h)$. Since $h = h(t)$ is a decreasing function on $[0, t_c]$, the inverse function is well defined on $[0, h_0]$. The time t corresponding to the height h is

$$t = h^\leftarrow(h) = \left[\frac{2(h - h_0)}{g}\right]^{1/2}.$$

The composite function $v = v(h^\leftarrow(h))$ associates a velocity v with each height h,

$$v = v(h^\leftarrow(h)) = v\left(\left[\frac{2(h - h_0)}{g}\right]^{1/2}\right) = -[2\,|g|\,(h_0 - h)]^{1/2}.$$

EXERCISES [1.5.1]

. . . .

1. Write the solution function of the equation $\dot{v} = g$ and initial condition $v(0) = v_0$, and the corresponding height function for $h(0) = h_0$.

2. Consider a 10-g weight which is projected upward with an initial velocity of 1000 cm/sec. If the acceleration units are cm/sec^2, a reasonable approximation of $|g|$ near the earth's surface is 980. Use the model based on constant acceleration to predict:

   a) the distance from the starting point at $\frac{1}{2}$, 1, 2 sec,
   b) the highest point reached and the time required,
   c) the total distance traveled in the first second and the first two seconds,
   d) (a), (b), and (c) for a 1-g object.

3. For a ball thrown upward with initial velocity v_0, compare the time required to:

   a) reach the highest point,
   b) return to the starting point.

. . . .

Observed results are consistent with these predictions if the object is sufficiently dense (a feather is not) and h_0 is small. For large h_0, the velocity increases enough to induce significant air resistance. Moreover the acceleration cannot be considered constant over a large range of heights. Other possible effects such as winds, nearby explosions, heavy magnets, etc., have also been ignored. Can models with fewer limitations be constructed? Can gravity and air resistance and other forces be taken into account simultaneously so that realistic predictions can be made in more complicated situations?

1.6 NEWTON'S LAW $F = ma$

Mechanics is the study of the motion of bodies subjected to certain forces. The basic law, formulated by Galileo and more precisely by Newton (1642-1727), states that an object moves with constant velocity unless a force acts on it. Re-phrased, the acceleration of a body indicates that a force is acting. Or, by definition, a *force* is that which accelerates an object. This concept of force is more precise and comprehensive than the notion of a push or pull. Gravitational and frictional forces confused the early searches for a general principle because no pushing or pulling agents were evident.

Another of Newton's laws, $F = ma$, provides a quantitive measure of force. In words, the acceleration, or change in velocity of an object, is proportional to the force applied. But the same push applied to a book or a piano gives different results. The piano offers greater resistance. The measure of an object's resistance to accel-eration is called its *inertial mass*. This mass may be defined as F/a, the ratio of the force to the change in velocity. For a given force, the larger the mass of the object, the smaller the resulting acceleration. These proportionalities have been established experimentally. Frictional forces are minimized by mounting objects on dry ice, or using an air table. Then a constant force F_c is applied over a time interval, and multiple-exposure photographs are taken as the object moves. Changes in velocity are computed from the sequence of photographs. The data fit the relation

$$F_c(t_2 - t_1) = m(v_2 - v_1)$$

to within expected experimental errors. Hence this relation is a unifying concept, or law, which "explains" the motion, and is useful in making predictions. The units may be chosen so that the factor of proportionality, between force and acceleration, is precisely the mass. For example, if the mass unit, called a *kilogram*, is defined as the mass of a particular cylinder of platinum alloy at Sèvres (France), and the force unit, a *newton*, is defined as the force which increases the velocity of a 1-kg object by 1 m/sec^2, then the ratio

$$F_c \frac{t_2 - t_1}{v_2 - v_1}$$

is the mass number in kilograms. Finally, in order to consider a variable force $F(t)$ as a function of time, let the instantaneous force be

$$F(t_1) = \lim_{t_2 \to t_1} m \frac{v_2 - v_1}{t_2 - t_1} = ma(t_1).$$

Excellent discussions of forces and Newton's laws are found in:

Feynman, R. P., R. B. Leighton, and M. Sands, *The Feynman Lectures on Physics*, Vol. 1, Chapters 9 and 12. Reading, Mass.: Addison-Wesley, 1965.

Holton, G., *Introduction to Concepts and Theories in Physical Science*, Chapter 4. Reading, Mass.: Addison-Wesley, 1952.

Physical Science Study Committee, *Physics*, 1st edition, Chapters 20-22. Boston: D. C. Heath, 1960.

EXERCISES [1.6.1]

. . . .

1. Consider a particle of unit mass at rest. A unit force pushes the particle to the right for $\frac{1}{2}$ min. Then a unit force pushes it to the left for the next $\frac{1}{2}$ min. These forces continue to alternate every half minute.

   a) Obtain and sketch the acceleration function, the velocity function, and the position function, for the first 2 min. It will be necessary to use closed intervals here, as in the model of the falling object.

   b) Where is the particle at the end of 10 min?

. . . .

The inertial-mass numbers are additive. That is, the function which associates with each object Q a mass number $m(Q)$ is a linear function. For example, the mass number of two objects attached together is the sum of their individual masses;

$$m(Q_1 + Q_2) = m(Q_1) + m(Q_2).$$

And the mass of four identical objects considered as one piece, is four times the mass of one of the objects, or, in general,

$$m(kQ) = km(Q).$$

Another concept of mass arose in the attempt to quantify a different situation. Every object is subject to a "pull" toward the earth. Two objects are said to have the same *gravitational mass* if they balance each other on a simple equal-arm balance. Mass numbers assigned to an object indicate the number of mass units needed to balance it. Hence the gravitational mass is the measure of an object's balancing power. This set of numbers is also additive. Moreover, for each object, the inertial-mass number and the gravitational-mass number are proportional (within the limits of our measuring techniques). Thus, a common unit, the kilogram, allows the use of the simple term *mass*. And the mass of an object may be determined on a simple balance or as a relative resistance to acceleration.

The magnitude of the earth's "pull," or gravitational force, on an object is commonly known as the weight of that object. Weight can be measured by hanging the object on a spring calibrated in newtons. At the surface of the earth a 1-kg object weighs approximately 9.8 newtons (N). At a fixed place, the weight of an object is proportional to its mass; the weight is $F_g = m|g|$, where g is the constant of proportionality. This factor $|g|$, the gravitational force per unit mass, differs from place to place. In New York it is 9.803 (N/kg); at the Canal Zone, 9.782. At each position outside the earth, consider a vector of magnitude $|g|$ (where $|g|$ is the proportionality constant at that position) directed toward the center of the earth. The collection of points and associated vectors makes up the earth's gravitational field, a three-dimensional analog of a direction field.

If an object in the earth's gravitational field is free to move, it falls to the earth. According to Newton's law, the force exerted, $F_g = mg$, must equal the product of the inertial mass and the acceleration of the object. Hence $mg = ma$, or $g = a$. In words: At a given position, all objects experience the same acceleration if forces other than gravity are negligible. Thus the proportionality constant g is often called the acceleration due to gravity and expressed in acceleration units. This special application of Newton's law explains Galileo's observation that bodies fall with constant acceleration. The acceleration remains constant throughout the fall if the distance is sufficiently small.

Newton's law is actually a vector law: F represents the net force, that is, the vector sum of the forces acting on the object. Hence it provides a method for considering several forces acting simultaneously.

1.7 FALLING BODIES SUBJECT TO RESISTING FORCES

A second model of a falling object includes a resisting force F_r as well as the gravitational force F_g. In the lower atmosphere, assuming constant air density and a restricted range of velocities, F_r is roughly proportional to the velocity of the object. The proportionality constant k, called the *drag coefficient*, is a property of the object dependent on its shape, size, and position, but not its mass. The drag coefficient is a measure of the resistance presented to the medium in which the object moves. Airplanes, ships, and rockets are designed to minimize drag. The techniques of calculating the drag for an arbitrary surface, over a wide range of velocities, are still being developed. Wind tunnels and water tunnels are used to determine the drag of specific objects and to accumulate the experience that must precede the theory.

The gravitational force, still considered constant, is directed toward the earth's center. The resisting force acts in a direction opposing the velocity, that is, $F_r = -kv$. Therefore, the net force acting on the object, at any instant of fall, is

$$F = F_g + F_r = mg - kv.$$

And, by Newton's law,

$$ma = mg - kv.$$

Hence the acceleration is no longer constant, but varies linearly as the velocity: $a = g - (k/m)v$. The mathematical model is the differential equation

$$\dot{v} = g - (k/m)v.$$

As before, the initial condition $v(0) = v_0$ specifies a velocity at $t = 0$, hence the time interval is extended to $[0, t_c]$.

Theorem 1.3.1 is of no help here; the velocity function is not specified by the relation

$$v(t) = \int_0^t g - \left(\frac{k}{m}\right) v(u) \, du + v_0.$$

The solution function cannot be defined in terms of itself. In the final chapter just such an integral equation plays a central role in generating approximate and numerical solutions. But the equation $\dot{v} = g - (k/m)v$ yields to simpler techniques. Before trying other methods, consider some predictions of motion implied by the differential equation itself.

Case 1: If v_0 is such that $g - (k/m)v_0 = 0$, then the initial acceleration is zero. Hence the velocity remains the same; the acceleration, $g - (k/m)v$, remains zero; and the object falls with constant velocity $v(t) = gm/k$. In this case, the initial condition $v_0 = gm/k$ restricts the differential equation to the simple case $\dot{v} = 0$. The equations of motion, then, by Theorem 1.3.1, are

$$a = 0, \qquad v = \frac{gm}{k},$$

and

$$h = \left(\frac{gm}{k}\right) t + h_0, \qquad t \in [0, t_c].$$

Case 2: If $g - (k/m)v_0 < 0$, the initial acceleration is negative (in the direction of fall) and the velocity function decreases (see Fig. 1.4). Since $v_0 > mg/k$, the

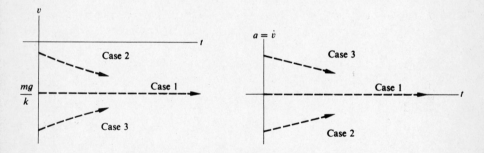

Figure 1.4

values $v(t)$ approach mg/k from above as t increases. The acceleration

$$\dot{v} = \frac{k}{m}\left(\frac{gm}{k} - v\right),$$

then, becomes smaller in magnitude; that is, $\dot{v}(t)$ increases, but not beyond zero. For, if $\dot{v} = 0$, the velocity remains constant thereafter, as in Case 1. Hence the velocity is bounded from below by the number mg/k.

Case 3: The reader may predict the motion where $g - (k/m)v_0 > 0$. (See Exercise 1 [1.7.1].)

Since $g - (k/m)v_0$ is a real number, it must fall in one of the three cases, that is, it is either equal to, less than, or greater than zero. Thus, for any initial velocity, the values $v(t)$ are bounded by the constant mg/k. But it is not clear whether $v(t)$ approaches mg/k as a limit in the true sense. For more precise predictions, the usual equations of motion that specify the position, velocity, and acceleration at any time t are desirable.

Can a function $v = v(t)$ on $I = [0, t_c]$ be obtained as a solution of the equation $\dot{v}(t) = g - (k/m)v$? The values $g - (k/m)v$ specify a function only if considered on a domain of velocity values. And Theorem 1.3.1 applies only to an equation of the form $y'(x) = f(x)$, where the solution function and its derivative have the same domain as f. Can the condition on $\dot{v}$ be restated in a differential equation that does yield to Theorem 1.3.1? The inverse function $v^{\leftarrow}(v) = t$, if it exists on $v[I]$, has a domain of velocity values. Might it be determined by Theorem 1.3.1?

The key here is the inverse function theorem from calculus: If f is differentiable and $f'(x) \neq 0$ on I, then $f^{\leftarrow}$ exists on $f[I]$ and $f^{\leftarrow\prime}(f(x)) = 1/f'(x)$. (See Exercise 2 [1.7.1].)

In particular, consider Case 2, where $g - (k/m)v_0 < 0$. Let

$$J = \left\{ v : v > \frac{mg}{k} \right\}.$$

Then for $v \in J$,

$$\dot{v} = g - \left(\frac{k}{m}\right) v < 0;$$

the velocity function is decreasing, and the inverse function exists on J. If $v = v(t)$ is differentiable, then so is the inverse function $v^{\leftarrow}(v) = t$, and

$$v^{\leftarrow\prime}(v) = \frac{1}{g - (k/m)v}.$$

This restatement of the condition on $\dot{v}$ is a differential equation of the form designated in Theorem 1.3.1. Moreover, the hypothesis of the theorem is satisfied, because the reciprocal of a continuous nonzero function is continuous. The unique solution of this differential equation and initial condition $v^{\leftarrow}(v_0) = 0$ is

$$v^{\leftarrow}(v) = \int_{v_0}^{v} \frac{1}{g - (k/m)u} \, du + 0, \qquad v \in J.$$

Can the velocity function be defined explicitly? In this case, the integral can be evaluated in terms of a simple antiderivative.

$$v^{\leftarrow}(v) = t = -\frac{m}{k} \left[\ln \left(g - \left(\frac{k}{m}\right) v \right) - \ln \left(g - \left(\frac{k}{m}\right) v_0 \right) \right].$$

The exponential function is a 1–1 mapping of R into R^+, that is, $a = b$ if and only if $e^a = e^b$. Thus

$$e^{-(k/m)t} = \frac{g - (k/m)v}{g - (k/m)v_0}.$$

Finally, solving for v, the velocity function may be specified as

$$v(t) = \frac{mg}{k} - \left(\frac{mg}{k} - v_0\right) e^{-(k/m)t}, \qquad t \in v^{\leftarrow}[J]. \qquad (1.7.1)$$

The second term is a constant multiple of an exponential function whose values approach zero as $t \to \infty$. Hence the velocity does approach the constant mg/k, called the *terminal velocity*. Since $g - (k/m)v_0 < 0$, the second term is positive and the velocity decreases toward mg/k. Because the exponential is a positive function, that is, it never assumes the value zero, the terminal velocity is never attained. Thus, for all $t \geq 0$, $v > mg/k$, and hence $J \supset v[I]$ and $v^{\leftarrow}[J] \supset I$; the solution is defined on $[0, t_c]$.

Practically, within the accuracy of measurement, the terminal velocity is often reached. For example, raindrops falling with zero initial velocity (even from heights of up to 1 km) effectively reach their terminal velocity after falling several meters. The velocity of hailstones is much greater (see Exercise 7 [1.7.1]); single stones may damage life and property. The terminal velocity of the human body is about 60 m/sec (120 mph).* Before opening his parachute, an experienced jumper can vary his fall velocity from 120 mph to 200 mph by changing his body position and thus decreasing the drag. A parachute presents a larger surface, about 6 m in diameter, and increases the drag coefficient k so that the terminal velocity is within survival range.

Examples of Case 3, where the initial velocity is of large magnitude and the object is slowed down to its terminal velocity, include objects entering the earth's atmosphere. It is the drag in this atmosphere that prevents excessive damage to the earth from meteorites, and provides the braking of manned capsules. On January 3, 1970, a meteor, photographed by a camera with a periodically interrupted time exposure at a station of the Smithsonian Prairie Network, fell at Lost City, Oklahoma, and became known as the Lost City meteorite. It entered the atmosphere (became luminous and hence a meteor) at 14 km/sec. At 25 km above the earth, the speed was 10 km/sec, and 3.5 km/sec at 19.5 km, where it ceased to be visible. Most impact velocities of meteorites have been estimated between 0.1 and 0.2 km/sec. This rapid deceleration is one of the hazards of the reentry of manned capsules. The drag coefficient of this 9.8-kg stone was estimated as 1. This allows an estimate

---

*1 m/sec $\doteq$ 2 mi/hr is a convenient intuitive aid in our nonmetric society. See Lord Ritchie-Calder, "Conversion to the Metric System," *Scientific American*, **223**, 1, 17–25, July 1970.

of the terminal velocity. The terminal velocities of meteors that are not completely vaporized are too small to produce light; in the last seconds of fall the object is a dark body.

Meteors also illustrate the limitations of this second model. At speeds greater than that of sound (0.3 km/sec), the resisting force varies as the velocity squared. During the visible part of the fall, the mathematical model of the meteor's motion is

$$m\dot{v} = -\tfrac{1}{2}\, c\rho A v^2,$$

where ρ is the atmospheric density, A the frontal area, and c^* the constant factor in the drag coefficient. For such high speeds, the value of g is insignificant relative to this term. During a fall of such great distance, ρ and A may change and cannot be incorporated in a constant of proportionality. Thus the model is not a simple equation, but a program of numerical methods in which values based on empirical data are introduced as the situation changes. A model with resisting force $F_r = -kv$ is reasonable only for small velocities; for the meteor, it applies only in the final seconds of fall.

In general, the drag depends on the object's shape, size, position, and velocity, as well as the density and viscosity of the medium. Even the rules for calculating the drag change, as these quantities change. At present, the ratio

$$\frac{\text{speed} \times \text{length} \times \text{density}}{\text{viscosity}},$$

called the *Reynolds number*, gives some indication as to how to predict the drag. The model $F_r = -kv$ is reasonable for a certain range of Reynolds numbers. For larger Reynolds numbers, the drag is proportional to powers of v greater than 1, and k may depend on surface area rather than the linear dimensions of the object. Neither units nor ranges of Reynolds numbers are specified here. The concept is mentioned only to indicate the present difficulties in constructing theoretical models which incorporate drag. For a lucid discussion of drag see Shapiro's *Shape and Flow* listed below.

SUGGESTED READINGS

Blanchard, D. C., *From Raindrops to Volcanoes*, Anchor s50.   Garden City, N.Y.: Doubleday, 1967.

Fireman, Edward L., "The Lost City Meteorite," *Sky and Telecsope*, **39**, 3, 158, March 1970.

---

*Note that c is less comprehensive than the coefficient k used in the simpler model of a falling body in the lower atmosphere.

Heide, Fritz, *Meteorites*, Phoenix Science Series 522. Chicago: University of Chicago Press, 2nd edition, 1964. Translation of *Kleine Meteoritenkunde*, 2nd edition. Berlin: Springer Verlag, 1957.

McCrosky, Richard E., "The Lost City Meteorite Fall," *Sky and Telescope*, **39**, 3, 154-158, March 1970.

Shapiro, A. H., *Shape and Flow*, Anchor s21. Garden City, N.Y.: Doubleday, 1961.

EXERCISES [1.7.1]

. . . .

1. Consider the equation $\dot{v} = g - (k/m)v$ in Case 3, where $v_0 < mg/k$.
   a) Predict the motion from the differential equation.
   b) Determine, if possible, an interval J on which $v^{\leftarrow}(v) = t$ exists.
   c) Define the solution $v = v(t)$ on I explicitly and compare it with the solution (1.7.1) in Case 2.
   d) Is a similar procedure possible or necessary in Case 1?

2. a) What conditions on f ensure the existence of $f^{\leftarrow}$?
   b) An intuitive justification of the inverse function theorem is suggested by sketching a decreasing differentiable (smooth) curve on transparent paper. Rotate the paper so that the positive x- and y-axes are interchanged. Is the smoothness affected? Compare the slope at (x, y) with the slope at (y, x) after rotation.

*   (c Show that

$$\lim_{x \to x_0} \frac{y - y_0}{x - x_0} = \frac{1}{\displaystyle\lim_{x \to x_0} \frac{x - x_0}{y - y_0}}.$$

One proof hinges on the argument that under suitable conditions

$$\lim_{x \to x_0} \frac{x - x_0}{y - y_0} = \lim_{y \to y_0} \frac{x - x_0}{y - y_0}$$

3. Let $k = 1$, $|g| = 10$, and $m = 1$ where the units are meters, seconds, and kilograms.
   a) Sketch the direction field of $\dot{v} = g - (k/m)v$.
   b) Sketch solution curves for $v_0 = -20, -5, 0$, and 5.
   c) The solution (1.7.1) may appear complicated, but actually consists largely of constants:

   $$v(t) = k_1 - k_2 e^{-k_3 t}.$$

   Obtain a solution of the equation

   $$\dot{h} = k_1 - k_2 e^{-k_3 t}$$

   and initial condition $h(0) = h_0$.
   d) Sketch the height function $h = h(t)$ for various initial heights and velocities.

4. Compare the velocities attained by a $\frac{1}{2}$-g object ($k = 0.6$ where the mass unit is 1 g) dropped from a five-story building, a ten-story building, and the Empire State building. [Note that in this model, k and m always appear as a ratio. The values of k are often adjusted to the mass units.]

5. A spherical raindrop 1 mm in diameter has drag coefficient $k = 1.25 \, (10)^{-3}$ if mass units are grams. Let $g = 9.8$ m/sec^2. If the drop falls from a height of 600 m with zero initial velocity,

   a) what is the terminal velocity?

   b) when is the velocity within 5% and 1% of the terminal velocity? What are the corresponding heights?

6. For objects of similar shape, k is proportional to the size of the object. In particular, for spherical raindrops, k varies as the diameter of the drop. If a hailstone maintains a spherical shape by adding concentric layers, what is the terminal velocity of a hailstone 0.7 cm in diameter?

7. Note the data given below.

PRECIPITATION VALUES*

	Diameter of drop	Terminal velocity	Height of cloud above surface
Fog	.01 mm	.003 m/sec	0 meters
Mist	.1	.25	100
Drizzle	.2	.75	200
Light rain	.45	2.00	600
Moderate rain	1.0	4.00	600
Heavy rain	1.5	5.00	1000
Excessive rain	2.1	6.00	1200
Cloudburst	3.0	7.00	1200
. . . .	5.0	8.00	

*W. J. Humphreys, *Physics of the Air*, 3rd ed., p.280. New York: McGraw-Hill, 1940.

For what range of diameters is k roughly proportional to the diameter of the drop? [The discrepancy for larger drops is due partly to a change in the Reynolds number, but also to a change in shape. As the diameter increases above 1 mm, the shape of the drop becomes more like a hamburger bun, and the flat lower surface presents a greater resistance. See McDonald, J. E., "The Shape of Raindrops", *Scientific American*, **190**, 2, 64, Feb. 1954.]

8. A parachutist, weighing 100 kg (including parachute) drops from a height of 2 km and opens the parachute after 3 sec. Predict his vertical motion. The value of k for a man with folded parachute is about 16 relative to kilogram mass units. The open parachute increases k by a factor of 10.

. . . .

1.8 THE EQUATION $y' = q(y)$

The differential equation $\dot{v} = g - (k/m)v$ was solved by reducing it to an equation already solved. In particular, the condition was replaced by an equivalent condition satisfying Theorem 1.3.1. This technique is common in mathematics, and indeed, in all types of problem-solving. Must this procedure be repeated whenever a similar type of equation arises? Or is it possible to apply the same method to the general equation $y' = q(y)$, where q is a specified function on the range of the desired solution function? A second theorem, applicable directly to any equation of this type would be convenient.

Consider the differential equation $y'(x) = q(y)$. A solution function $s: x \to y$ and its derivative $s': x \to y'$ have the same domain I. The function q is specified on $J = \{y: y = y(x), x \in I\}$. The inverse function $s^{\leftarrow}$ or $y^{\leftarrow}(y) = x$, if it exists on J, has the derivative $y^{\leftarrow\prime}(y) = 1/q(y)$. This last equation yields to Theorem 1.3.1. But can the inverse function be used in general, as it was in the previous section? What conditions ensure the existence of the inverse? (See Exercise 2 [1.7.1].) If the solution s is increasing (or decreasing), or equivalently, if the derivative s' is positive (or negative) on I, then the inverse $y^{\leftarrow}(y) = x$ exists on J. This last condition may be imposed on the function q. Thus, if q is positive on J, so is $y' = q(y)$ and $y^{\leftarrow}(y) = x$ is defined on J. And by the inverse function theorem, $y^{\leftarrow\prime}(y) = 1/q(y)$.

Finally, the hypothesis of Theorem 1.3.1 is satisfied if the reciprocal function $1/q: y \to 1/q(y)$ is continuous on J, that is, q must be continuous and nonzero on J. This implies that q must be always positive (or negative). For the intermediate-value theorem of calculus states that if f is continuous on an interval and $f(x)$ assumes the values a and b, then $f(x)$ assumes all values between a and b. Thus if f is continuous and $f(x)$ assumes positive and negative values, $f(x) = 0$ for some x in the interval.

Hence the condition that q be continuous and nonzero on J is sufficient to ensure that $y^{\leftarrow}(y) = x$ exists on J and can be written as an integral function. This suggests the following theorem.

Theorem 1.8.1 [on solutions of $y' = q(y)$]

Let q be a continuous, positive (or negative) function on the interval J.

Let (x_0, y_0) be any point such that $y_0 \in J$.

Then there is a unique function, on an interval I with range J, satisfying the differential equation $y' = q(y)$ and the condition $y(x_0) = y_0$.

This solution is defined as the inverse of the function

$$y^{\leftarrow}(y) = x = \int_{y_0}^{y} \frac{1}{q(u)}\, du + x_0, \qquad y \in J.$$

Both theorems have been preceded by a line of reasoning intended to motivate and justify the theorem. But, in general, the opposite order, a statement of the theorem followed by a proof, has advantages. It is easier to be brief and precise once the theorem is stated, and the reader knows where the argument is leading. In spite of the duplication, a proof for Theorem 1.8.1 will be given. It is suggested that the reader write his own proof before proceeding to the one below. The remarks preceding the statement of the theorem may be useful in framing a proof.

Proof: If q is positive, continuous, and nonzero, then $1/q(y)$ is positive on J. Thus the integral function

$$x = y^{\leftarrow}(y) = \int_{y_0}^{y} \frac{1}{q(u)} \, du + x_0 \tag{1.8.1}$$

is defined on J and has a positive derivative $y^{\leftarrow\prime}(y) = 1/q(y)$. Hence the inverse function, call it $s: x \to y$, exists and $s'(x) = y' = q(y)$. That is, s satisfies the differential equation.

The intermediate-value theorem and the fact that the integral function (1.8.1) is increasing imply that $I = s^{\leftarrow}[J]$ is an interval. Since

$$y^{\leftarrow}(y_0) = \int_{y_0}^{y_0} \frac{1}{q(u)} \, du + x_0,$$

it follows that $s: x_0 \to y_0$.

Finally, the solution is unique. Let w be any function on an interval I_a containing x_0 such that $w'(x) = q(y)$ and $w(x_0) = y_0$. The conditions on q imply that w is an increasing function and its inverse is an antiderivative of $1/q$. But there is only one such antiderivative whose value at y_0 is x_0. Hence $w^{\leftarrow}(y) = s^{\leftarrow}(y)$ on the common interval containing y_0. Since the inverse function is unique, it follows that $w(x) = s(x) = y(x)$ on $I \cap I_a$. ∎

It makes sense to talk about the maximum interval of a solution function. The endpoints of each interval containing x_0 are real numbers. Consider the set of admissable intervals for a particular function. The set of right endpoints has a least upper bound (lub), and the set of left endpoints has a greatest lower bound (glb), which may be ∞ or $-\infty$ respectively.

If $s: x \to y$ is a solution on (x_0, y_0), then for each x in I relation (1.8.1) holds. The endpoints of the maximum open interval (α, β) for this solution are

$$\alpha = \operatorname*{glb}_{y \in J} \int_{y_0}^{y} \frac{1}{q(u)} \, du + x_0 \quad \text{and} \quad \beta = \operatorname*{lub}_{y \in J} \int_{y_0}^{y} \frac{1}{q(u)} \, du + x_0.$$

If the integral function can be expressed in terms of elementary functions, then the solution may be determined explicitly by solving for $y(x)$ in terms of $y^{\leftarrow}(y) = x$.

In any case, $s: x \rightarrow y$ is well defined and can be approximated by numerical methods.

EXERCISES [1.8.1]

....

1. Consider the equation $y' = y^2 + 1$.
   a) Sketch the direction field.
   b) Determine a maximum interval J for which the hypothesis of Theorem 1.8.1 is satisfied.
   c) Find the solution curve through $(0, 0)$ and check it with the direction field. Determine the largest domain for this solution. Can it be extended to the closed interval?
   d) Determine solutions through $(\pi/4, 0)$ and $(0, 1)$. In each case state the maximum domain.
* e) Write an expression for the maximum domain of the solution through (x_0, y_0) in terms of the coordinates x_0 and y_0.

2. Consider the equation $y' = \tan y$.
   a) Determine a solution through $(\ln \frac{1}{2}, \pi/6)$ and check it with the direction field.
* b) Can the solution be extended to include $(0, \pi/2)$?

3. Consider the equations
   i) $y' = y$,   ii) $y' = y^2$,
   iii) $y' = 1/y$,   iv) $y' = 1/y^2$

   a) Sketch the direction fields and obtain solutions (on maximum domains) through $(1, 1)$ and $(0, 1)$.
* b) Can the domain, in each case, be extended to the endpoints of the interval?
* c) Can these solutions be extended to include a point $(x, 0)$?

*4. Consider a falling raindrop with the conditions in Exercise 5 [1.7.1]. Use the mathematical model $\dot{v} = g - (k/m)v$ to set up procedures for numerical approximations on the time interval $[0, 10]$. Try methods based on the differential equation and Simpson's rule. Carry out enough steps to check your procedures.

....

*1.9 UNIQUENESS

Consider $\dot{v} = g - (k/m)v$, the second model of a falling object, as a special case of $y' = q(y)$. To satisfy the hypothesis of Theorem 1.8.1, the domain of $q(v) = g - (k/m)v$ must be limited to values of $v > mg/k$ (or $< mg/k$). In each of the corresponding regions in the tv-plane, there is a unique solution through each point. But if the conditions are ignored and q is considered on R, there is still a

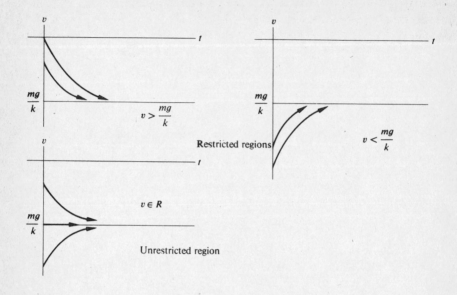

Figure 1.5

unique solution through each point in the plane (see Fig. 1.5). And all the solutions make sense physically. In short,

$$v(t) = \frac{mg}{k} - \left(\frac{mg}{k} - v_0\right) e^{-k(t-t_0)/m}, \qquad t \in R,$$

specifies a unique solution through each point in the plane. Is it wasted effort to be concerned about domains of the functions involved? What is the value of these restrictions? Are they just devices to allow mathematicians to prove theorems? These questions are best answered by another example.

Consider the equation $y' = |y|^{1/2}$. In order that $q(y) = |y|^{1/2}$ be unequal to zero, restrict the domain to R^+. For $y > 0$, the equation may be written $y' = y^{1/2}$. By Theorem 1.8.1, the unique solution satisfying $y(x_0) = y_0$ has an inverse defined explicitly as

$$y^{\leftarrow}(y) = x = \int_{y_0}^{y} \frac{1}{q(u)} \, du + x_0 = 2y^{1/2} - 2y_0^{1/2} + x_0, \qquad y \in R^+.$$

The range of this inverse function is the set $\{x: x > x_0 - 2y_0^{1/2}\}$. Hence the solution itself is specified as

$$y(x) = \tfrac{1}{4}(x - (x_0 - 2y_0^{1/2}))^2, \qquad x > x_0 - 2y_0^{1/2}.$$

The graph $\{(x, y): y = \tfrac{1}{4}(x - (x_0 - 2y_0^{1/2}))^2, x \in R\}$ is a parabola with vertex

$(x_0 - 2y_0^{1/2}, 0)$. Hence the solution curve is the right half of this parabola. Solution curves through points in the upper half-plane are shown in Fig. 1.6(a).

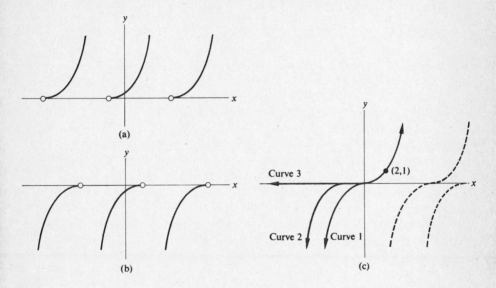

Figure 1.6

Similarly, if the domain of q is R^-, the equation may be written $y' = (-y)^{1/2}$ and the unique solution on (x_0, y_0) is specified as

$$y(x) = -\tfrac{1}{4}(x - (x_0 + 2(-y_0)^{1/2}))^2, \qquad x < x_0 + 2(-y_0)^{1/2}.$$

Solution curves in the lower half-plane are the half parabolas in Fig. 1.6(b). For the domain $y = 0$, the zero function on R is a solution on $(x_0, 0)$. Its curve is the x-axis.

Consider q on R and sketch the solutions on a single set of axes. Then, through any point in the plane, there are many solution curves. Three solutions on $(2, 1)$ are shown in Fig. 1.6(c). These curves may be defined explicitly as follows:

$$\text{Curve 1} \begin{cases} (x, y): y = \tfrac{1}{4}x^2 & \text{if } x \geq 0, \\ \quad\quad\;\; y = -\tfrac{1}{4}x^2 & \text{if } x \leq 0; \end{cases}$$

$$\text{Curve 2} \begin{cases} (x, y): y = \tfrac{1}{4}x^2 & \text{if } x \geq 0, \\ \quad\quad\;\; y = 0 & \text{if } -1 \leq x \leq 0, \\ \quad\quad\;\; y = -\tfrac{1}{4}(x + 1)^2 & \text{if } x \leq -1; \end{cases}$$

$$\text{Curve 3} \begin{cases} (x, y): y = \tfrac{1}{4}x^2 & \text{if } x \geq 0, \\ \quad\quad\;\; y = 0 & \text{if } x \leq 0. \end{cases}$$

Many more solution curves through this point may be sketched and specified. This example demonstrates that if the domain is not restricted as in Theorem 1.8.1, a solution curve through a given point may not be unique. The physical interpretation is that at (2, 1) the situation is not predictable; that is, under the same conditions different types of motion might occur.

In each restricted domain R^+, R^-, and 0, there is a unique solution through each point in the domain and a general form for specifying the inverse function. In the unrestricted domain, not only is the uniqueness lost, but there is no longer a single method or form for specifying all solutions. In short, the mathematical situation in the unrestricted domain may be chaotic.

The proof of Theorem 1.8.1 indicates that the hypothesis is sufficient to imply the conclusion. The equation $\dot{v} = g - (k/m)v$ indicates that the hypothesis is not necessary; that is, the conclusion may hold even if the hypothesis does not. Equation $y' = y^{1/2}$ indicates that the hypothesis is not too strong; that is, if it does not hold, the conclusion may not hold.

The restricted domain is often bounded by a number c such that $q(c) = 0$. It can be shown that the uniqueness of solutions in a domain including c depends on whether or not the improper integral $\int_{y_0}^{c} 1/q(u)\, du$ exists (see Exercises [1.9.1]).

EXERCISES [1.9.1]

1. Consider the equations $y' = y^{1/2}$ and $\dot{v} = g - (k/m)v$. The hypothesis of Theorem 1.8.1 is satisfied on the domains $\{y: y > 0\}$ and $\{v: v > mg/k\}$. Consider the larger domains $\{y: y \geq 0\}$ and $\{v: v \geq mg/k\}$. It has already been shown that through any point in the region $\{(t, v): v \geq mg/k\}$ there is a unique solution of $\dot{v} = g - (k/m)v$.

   a) Does the equation $y' = y^{1/2}$ have a unique solution through a point $(x_0, 0)$ in the region $\{(x, y): y \geq 0\}$?

   b) Are $y(x) = 0$ and $v(t) = mg/k$ solutions of the respective equations?

   c) Do the improper integrals

   $$\int_{v_0}^{mg/k} \left(g - \left(\frac{k}{m} \right) u \right)^{-1} du \quad \text{and} \quad \int_{y_0}^{0} u^{-1/2}\, du$$

   exist?

*2. Consider the general equation $y' = q(y)$. Assume q is continuous on $(a, b]$, and $q(b) = 0$.

   a) Show that $y(x) = b$, $x \in R$, is a solution curve.

b) Show that if $\int_{y_0}^{b} 1/q(u)\,du = c$, then any solution curve through (x_0, y_0) in the region $\{(x, y): y \in (a, b)\}$ may be extended to the point $(c + x_0, b)$.

c) Show that the combination of a solution curve through (x_0, y_0) in the region $\{(x, y): y \in (a, b]\}$, joined to a piece of the line $y = b$ at $(c + x_0, b)$, is a solution curve.

d) Specify graphically several possible solutions through a point (x_0, b) in the region $\{(x, y): y \in (a, b]\}$.

*3. Consider the case where $\int_{y_0}^{b} 1/q(u)\,du = \infty$. Describe the solution curves as $y \to b$. Do the solutions in the restricted region and the solution $y = b$ ever meet? Test your conclusion on the equation $y' = y^2$.

. . . .

1.10 THE MEASUREMENT OF TIME

A basic problem in measurement is to find a suitable unit of comparison. For time it must be a unit of duration. And unlike distance units which may be used over and over again (for example a straightedge or ruler), a unit of duration, once passed, is gone. The problem is to find a unit which repeats itself, that is, a periodic cycle that never changes. Or, to be more realistic, one whose variations are small compared to the accuracy desired. The earliest timekeeper seems to have been the earth—the unit was a complete revolution of the earth relative to the sun or the stars. Astronomers used the sidereal day, the time for a star to return to its observed position. Laymen used the solar day, the period between successive sunrises, or sunsets. Evidences of the last unit remain in our culture—for example, the word "fortnight" —and the evening festivities of the older religions whose days begin at sunset.

Later, probably in Babylonia, the period of light between sunrise and sunset was divided into twelve equal parts called hours. (The word "noon" derives from the nones, the office of the church to be said at the ninth hour but later moved to midday.) In the tropics such a division is fairly constant, but at greater latitudes, such hours vary considerably with the seasons. Even so, these hours continued to be used in Europe until the fourteenth century and came to be known as "temporary hours." The earliest instruments for measuring hours were the gnomon and sundial —evidently in use by 2000 B.C. (with refinements continuing through many centuries). But even in the third century B.C., the upheaval caused by their introduction into Rome is suggested by the following translation of a poem attributed to Plautus (254?–184 B.C.).*

———————
*Milham, W. I., *Time & Timekeepers*, pp.37-38. New York: Macmillan, 1923.

> The Gods confound the man who first found out
> How to distinguish hours—confound him, too
> Who in this place set up a sundial
> To cut and hack my days so wretchedly
> Into small pieces! When I was a boy
> My belly was my sundial—one more sure,
> Truer and more exact than any of them.
>
> The dial told me when t'was proper time
> To go to dinner, when I ought to eat;
> But, now-a-days, why even when I have
> I can't fall to unless the sun gives leave.
> The town's so full of these confounded dials.
> The greater part of its inhabitants
> Shrunk up with hunger, creep along the street.

The desire to limit the length of lectures arose even in early civilizations. In general, as life became more complex there developed a need to measure time intervals independently of the sun. The first mechanical device was the water clock—now known under the Greek name, *clepsydra*. The oldest specimen is an Egyptian water clock bearing the name Amenhetep III (dating from about 1400 B.C.). This particular clock was designed to count the night hours. Water basins and sandtimers, which are also in evidence among the early writings and artifacts, may be considered as primitive manifestations of the same principle. They all depend on the motion of a substance through an orifice.

SUGGESTED READINGS

Bartrum, C. O., "Time: Its Determination, Measurement, and Distribution," in *Splendour of the Heavens*, edited by T. E. R. Phillips and W. H. Steavenson. London: Hutchinson & Co., 1923.

Breasted, J. H., "The Beginnings of Time Measurement", in *Time & Its Mysteries*, Series I. New York: New York University Press, 1936.

Lyons, Harold, "Atomic Clocks," *Scientific American*, **196**, 2, 71–82, Feb. 1957.

Milham, W. I., *Time & Timekeepers*, pp. 37-38. New York: Macmillan, 1923.

Millikan, R. A., "Time," in *Time & Its Mysteries*, Series I. New York: New York University Press, 1936.

Noble, J. V., and deSolla Price, Derek J., "The Water Clock in the Tower of the Winds," *American Journal of Archaeology*, **72**, 345-355, Plates 111–118, 1968.

1.11 CLEPSYDRA

The clepsydra was the standard device for measuring time from at least 1400 B.C. to 1500 A.D. A simple example of a primitive clepsydra is an open cylindrical tank with a circular opening of radius r in the base (Fig. 1.7). If the tank is filled with

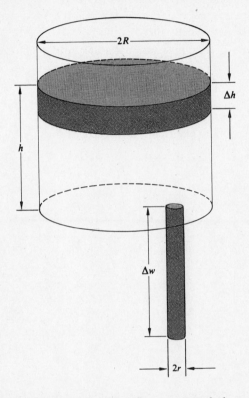

Figure 1.7

water and the opening unplugged, water flows out and the water level, h, falls. Is the height h a reliable indicator of time? Rephrased, does a fall of one unit always correspond to a fixed unit of time? Or, is the water level, considered as a function of time, a linear function? The question calls for a mathematical model that will determine a function specifying the water level h at any time t. The gradual change in water level suggests that it is reasonable to assume that the function $h = h(t)$ is differentiable for $t \geq t_0$, the time of unplugging.

Imagine a tube of radius r extending from the opening. In a time interval $\Delta t = t_2 - t_1$, let Δh and Δw denote the change in water level in the tank and tube, respectively. The water flowing out of the tank goes into the tube; that is,

$$\pi R^2\, \Delta h = \pi r^2\, \Delta w.$$

If $h = h(t)$ is differentiable, the limits of the average velocities of flow exist:

$$\lim_{\Delta t \to 0} \frac{\Delta h}{\Delta t} = \lim_{\Delta t \to 0} \frac{r^2}{R^2} \frac{\Delta w}{\Delta t},$$

or $\dot{h} = \dot{w} r^2 / R^2$. The values $\dot{w}(t)$ represent the particle velocity of the water, the rate at which the particles of water move through the orifice. Experimentally, the magnitude of the particle velocity through this type of opening has been determined as $0.6\sqrt{2|g|h}$,* where g is the gravitational acceleration. Thus

$$\dot{h} = \frac{-r^2}{R^2} 0.6\sqrt{2\,|g|}\,\sqrt{h}.$$

Note that h is a decreasing function, hence the negative sign. The implications of this mathematical model explain various aspects of the ancient clepsydras.

EXERCISES [1.11.1]

. . . .

Consider a cylindrical vessel 40 cm high, 30 cm in diameter, with an opening 1 mm in diameter in the base. Let the differential equation

$$\dot{h} = -\frac{r^2}{R^2}\left(0.6\sqrt{2\,|g|}\right) h^{1/2}$$

be the model representing the fall of the water level h.

1. Determine a region of the th-plane in which there is a unique solution curve through each point. Specify the solution function on a point (t_0, h_0) in this region. Indicate the maximum domain for this solution.

2. Indicate a larger region in which there is a solution curve, not necessarily unique, through each point. Indicate two specific curves through a particular point; specify the functions and sketch the curves.

3. Sketch the solution satisfying the initial condition $h_0 = 40$ cm.

Figure 1.8

_____

*Theoretically, the particle velocity of water falling from a height h is $\sqrt{2|g|h}$, as derived in Section 1.5. But the direction of flow through the orifice persists, the stream of water continues to contract, and the opening is, in effect, smaller (Fig. 1.8).

4. How long does it take for the full tank to empty? How long does it take a full tank to become half full?

5. Determine the maximum variation in the time interval required for a fixed change in the water level, say $\Delta h = 1$ cm. Might this variation have been observed 3500 years ago?

6. If hour marks on the inside of the vessel are uniformly spaced, determine the maximum variation in the hours so measured. Determine the spacing of lines to indicate uniform hours.

7. Make a clepsydra by punching a hole in the base of an open tin can. Test the predictions of the differential equation. Does water temperature affect the time intervals measured?

8. Discuss simple modifications in design that eliminate the nonlinear flow, or find examples of modified clepsydras.

. . . .

The following review and comments by the archeologist W. M. F. Petrie indicate the role that mathematical models can play in archeology.* One of the crucial dimensions, the size of the orifice, is not mentioned. It is probably difficult to ascertain once the vessel is broken. From the data given, can you estimate the size of the original openings?

In this paper a full account is given of two clepsydra vessels in the Cairo Museum, which were to be filled with water, and the lapse of time read off on graduations inside as the fluid dropped out. There were various fancy devices in the classical times for indicating the level by floats; whether such were used in these instances is immaterial, as the graduations inside show decisively how they were to be read.

Both vessels alike have twelve lines down the inside from top to bottom; and, as in one case, they have the names of the twelve months placed over them respectively, it is clear that a different upright line was read each month. Further, there are twelve lines around the side in one, presumably corresponding to twelve hours, and eleven lines with the top edge, making twelve, in the other; and these lines, while parallel at the top, become increasingly tilted going toward the bottom. Each vessel has a dropping hole near the bottom. Such are the main features in common to both clepsydras.

The earlier vessel is the more precise and interesting. In the section here copied from M. Daressy, the mean line through each row of points has been drawn outside of the vessel to demonstrate the increasing tilt on descending in the vessel. The bowl itself is of alabaster, 14.6 inches high outside, 12.6 inside. Top diameter 18.9 and 17.4; base diameter 10.2 out, 8.5 inside. It has been broken up, anciently, but most of the pieces have been recovered in the clearance of the ruins of Karnak by M. Legrain. The vessel bears figures of gods and inscriptions, naming Amenhetep III; this dates the bowl to 1380 B.C, or within thirty years before that. The vertical lines and hour lines around are marked by small drill holes.

*Petrie, W. M. F., "Review of Daressy, G., 'Deux Clepsydras Antiques' " (*Bull. Inst. Egypt*, V, IX, 1915), *Ancient Egypt*, pp.42-44, 1917.

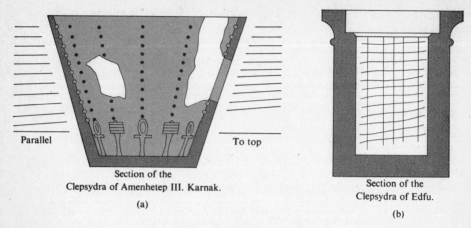

Parallel   To top

Section of the
Clepsydra of Amenhetep III. Karnak.

(a)

Section of the
Clepsydra of Edfu.

(b)

Figure 1.9

The signs at the bottom shown here mean life and stability, they are often used decoratively, and have no connection with the measurement.

How are we to understand this variation in the level along with the month of reading? M. Daressy concludes that the variation was to allow of the intervals corresponding to the longer or shorter hours of the night. But on comparing the lengths with the actual length of the night, it is shown that the variation of the long and short nights is nearly double the actual variation of the scales of summer and winter. We can go further by the marking of the names of the separate months. The first month was Thoth, and the largest scale here is of the month Hathor, beginning 17th September, and the shortest scale is of Pachons, which began 14th March in 1380 B.C. Now this is contrary to the theory, as the maximum and minimum are not at the solstices, as they should be. Evidently we must look elsewhere for an explanation.

A matter often overlooked—even in canal discharges—is the large variation in the viscosity of water with temperature. Much more will flow through a pipe as the temperature is raised; at boiling point the flow is six times the amount that it is when near freezing. If this difference in the greater length of the summer scale were to compensate for the quicker flow of the water, it would imply in the Karnak clepsydra a variation of 15 : 17 in the rate. At about 70° Fahr., the mean temperature of Egypt, this would imply a change of temperature of about 9° Fahr. between summer and winter. As these clepsydra were probably placed in the inner chambers of very massively built temples, with walls many feet thick, and close to the ground, it is likely that the change with the seasons might not exceed this.

Further, the maximum is in the middle of September, and the minimum in the middle of March, and those would be about the times of highest and lowest temperature, delayed by the massive building, and the conduction of heat through the earth. Thus the compensation for the viscosity of water agrees with the amount and the time of the change of scale. Probably it was empirically noticed that the rate of flow varied in summer and winter, and the scales were made to compensate this.

The Egyptian had gone further than this. It will be seen that the bowl has very sloping sides, and M. Daressy remarks this as an evidence of the merely ornamental nature of the work, as it would contain much more water between the upper than between the lower divisions. He concludes, therefore, that this work was fanciful, and was rendered useless by reason of ignorance. It is precisely this form which renders it more exact than a cylindrical vessel. Fluid drops faster in proportion to the pressure; when the clepsydra was full, it would drop nearly four times as fast as when the water had gone down to the lowest graduation. This change of rate the Egyptian tried to compensate by increasing the amount in the upper levels. Strictly, a parabolic outline is needed for a vessel with an equal scale down it. This was nearly attained by taking a frustum of a cone and not carrying the scale to the bottom. The variation from a parabola over the part used would not be large. On testing this, the rate of dropping, when the water was at the top or bottom of the scale, would vary as 1 : 3.7, and the ratio of the areas or quantity of water at those levels is 1 : 2.6. Further, the scale is rather more open at the top than below, which would make the ratio of quantities 1 : 2.9. Thus the sloping form of the bowl and variation of graduation compensate more than three quarters of the error due to the variable head of water. As the flow averaged nine drops a second, it must have been a stream at the beginning, ending in rapid drops.

It appears then that by 1380 B.C., the clepsydra was mainly compensated, both for the changes of water level and for the changes of temperature. The knowledge of this was kept up in Egypt, as Athenaeus, of Naukratis, in the third century A.D., says that water "which is used in hourglasses does not make the hours in winter the same as those in summer, but longer, for the flow is slower on account of the increased density of the water."

The second clepsydra is from Edfu, of apparently the Ptolemaic Age. It is a cylinder of hard limestone, with a cornice round the outside and an expansion of the cylinder at the top, probably to hold a lid. It is 14.9 inches high, 13.5 across at the top and 11.8 below. Inside, the graduated cylinder is 10.8 deep and 6.6 across. The vertical lines of the months are not equally spaced. The circumference seems to have been divided by halving into eighths; then two months on each side, at the maximum slope of the circular lines, have been subdivided. The effect would be to agree with a lengthening out of the mid-season period and a shortening of the hot and cold period. The difference between hot and cold season is rather greater than in the Karnak clepsydra, agreeing to a fluctuation from about 68° to 80° Fahr. There was thus the temperature compensation as in the earlier example; but being a cylinder, and having equal divisions, there was no compensation for the flow varying by the pressure. Owing to carrying down the divisions close to the bottom, this variation is the greater, and the flow at the beginning would be eighteen times quicker than at the end. It would be by drops, as they would average only two drops a second.

The monthly variation in the flow being then fully accounted for by the temperature compensation, it follows that the Egyptian used equal hours and did not lengthen or shorten them according to the seasons. This is what we might expect from the accurate astronomical observation of early date, proved by the precise agreement of the pyramids to the cardinal points. In this the Greek and Roman usage was more primitive, as they divided the day and night into twelve hours, of whatever length the season made them, as the Turk does at present. Hence adjustments were required to regulate the flow of clepsydras according to the length of the day, as described by Vitruvius . . .

A point of Egyptian usage should be noted. Although Mesore was reckoned the first month of the year in the XIIth dynasty, yet by 1380 B.C. Thoth was the first month, as the third month on the clepsydra has the figure of Hathor, fixing that month to be hers.

The Thoth year is, therefore, to be used in all reckonings of the XVIIIth dynasty, agreeing with its being more correct, astronomically, than the Mesore year from 1700 B.C. onward.

The Amenhetep III clepsydra was easily dated. But the well-preserved records of Egypt, including a calendar spanning 5000 years, are unusual. Before 1950, many archeological finds were dated only sequentially by the level at which they were found. How can undated objects from the past be located in our time scale?

1.12 CARBON-14 DATING

The chronology of ancient history and prehistory presents a difficult problem. Is it possible to measure time intervals extending back in time that predates preserved records or even the existence of man? Quite recently, it was realized that radioactive substances are clocks—capable of measuring times in the distant past. In particular, carbon-14 is used to date archeological finds. The methods and results are still being refined; the basic process is described here.

An isotope of an element contains in its nucleus the same number of protons as, but a different number of neutrons than, the element. The carbon nucleus has 6 protons and 6 neutrons (atomic number 6, atomic weight 12). The isotope carbon-14 (C^{14}) has 6 protons and 8 neutrons (atomic number 6, atomic weight 14). Carbon-14 is an unstable isotope; it is radioactive. Individual nuclei disintegrate. In particular, a neutron disintegrates; it leaves a proton and emits a β-particle (an electron moving at high speed). The nucleus then has 7 protons and 7 neutrons (atomic number 7, atomic weight 14) and is the stable nucleus of a nitrogen atom. The β-rays, together with α- and γ-rays emitted by isotopes of other elements, gave rise to the term "radioactive."

The radio-isotope C^{14} is produced continually by cosmic radiation. Cosmic-ray neutrons transmute nitrogen into C^{14}, which is uniformly distributed throughout our atmosphere in the form of carbon dioxide. Carbon dioxide is absorbed by plants which are in turn eaten by animals. Hence all living matter contains radioactive carbon.* In living tissues, the assimilated C^{14} atoms balance the disintegrating C^{14} atoms. At death, the assimilation process stops, but the disintegration process continues. Thus the amount of C^{14} in a sample of wood indicates the age of the wood. How is the amount of C^{14} measured? How does this measure determine the age of the sample?

Radioactive disintegration is a random process. The disintegration of a particular nucleus can be neither predicted nor controlled. Nor is it evident to the human senses. But in a Geiger counter, the β-particle emitted from a C^{14} nucleus

---

*It is interesting, if irrelevant, to note that the realization that all living matter contains radiocarbon produced by cosmic radiation resulted from researches on sewage gas in Baltimore. See W. F. Libby, E. C. Anderson, and J. R. Arnold, "Age Determination by Radiocarbon Content: Worldwide Assay of Natural Radiocarbon," *Science*, **109**, 227-228, March 4, 1949.

results in a pulse of current. This pulse may be amplified and recorded by an electronic device. But the counting of random events is not exact and repeatable as it is for events occurring at regular intervals. Nearly simultaneous radiations may not be distinguished. The radiations are also random in direction and kinetic energy. The β-particle may be directed into rather than out of the substance. Or it may not have enough kinetic energy to escape. The C^{14} atoms in the environs of the apparatus also contribute to the count. Hence the recorded counts are neither precise nor repeatable. However, the average number of counts over a suitable time interval do conform to a definite pattern. These average counts may be predicted and repeated within a certain range of error. This error depends on the equipment and its noise level (the average count with no sample present). The noise level can be minimized but not eliminated. The average count of emitted β-particles per time interval (above the noise level) determines the rate of disintegration of C^{14} in the sample.

Consider the amount A of C^{14} in a nonliving sample as a function of time. If the units are single atoms and microseconds, A decreases by one atom at irregular intervals and $A = A(t)$ is a step function (see Fig. 1.10a). If the units are grams and

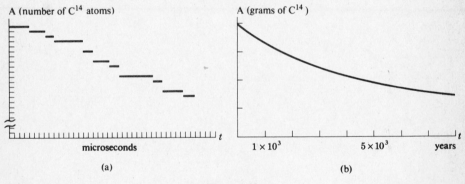

Figure 1.10

years, the steps are so small that the function curve appears to be continuous (see Fig. 1.10b). The phenomenon exhibits quantum effects in the small; but in the large, a smooth curve is a useful model and provides a method of archeological dating.

If $A = A(t)$ is assumed to be differentiable for $t > 0$, then the rate of disintegration may be represented by $\dot{A}(t)$.

All counts indicate that the rate of disintegration is proportional to the amount of C^{14} remaining, that is, $\dot{A} = -kA$. The derivative is negative because $A = A(t)$ is a decreasing function. The linear function $q(A) = -kA$ is nonzero and continuous for $A > 0$. According to Theorem 1.8.1, there is a unique function satisfying this equation and the initial condition $A(0) = A_0$. Actually, the inverse function, which defines the solution, is more useful here, since it determines the age t explicitly.

The differential equation is also a model for substances other than C^{14}, but the constant k differs for various elements. Usually k is determined by the half-life. Half the C^{14} atoms in any sample disintegrate in 5685* years (± 35 years), that is, $A(5685) = \frac{1}{2}A_0$. This is fortunate for archeology. Measured half-lives of other substances range from 10^{-12} seconds to 10^{12} years. Lead-210, for example, has a half-life of 22 years. Disintegration rates of white lead paint provide a method of dating paintings† and hence detecting forgeries within the last two centuries.

The average count of emitted β-particles indicates the rate $\dot{A}$ of disintegration. The differential equation $\dot{A} = -kA$ implies that

$$\frac{\dot{A}(t)}{\dot{A}(0)} = \frac{kA(t)}{kA_0} = \frac{A(t)}{A_0}.$$

Hence, for computations, the ratio of disintegration rates may be used directly in the solution function or its inverse. Consider the implications of this mathematical model in two actual cases described in the exercises below.

EXERCISES [1.12.1]

. . . .

1. Charcoal from the occupation level of the Lascaux cave in France gave an average count of 0.97 disintegrations/min/g over the noise level. Living wood gave a count of 6.68. Estimate the date of occupation, and hence the probable date of the remarkable paintings in the Lascaux cave.

2. In the 1950 excavations at Nippur, a city of Babylonia, charcoal from a roof beam gave a count of 4.09 (± 0.04) disintegrations/min/g. Dated tablets on the same building placed the construction between the first year of the reign of Shu-Sin and the third year of his successor Ibi-Sin. This sample is of particular interest because the Babylonian calendar has not been related to our own. Hypothesized dates for Hammurabi's accession, based on historical evidence, range from 2374 to 1975 B.C. Hence it is difficult to relate Mesopotamian civilization to contemporary cultures in Egypt and India.

   What is the likely time of Hammurabi's accession in view of the historical estimates and the radiocarbon dating? The statistical error in this count implies that there are 2 chances in 3 that the radiocarbon date is correct within 100 years, and 19 chances in 20 that the date is correct within 200 years.

. . . .

---

*Half-life accepted at Symposium on Radioactive Dating, Athens, 1962.

†Keisch, Bernard, "Dating Works of Art through their Natural Radioactivity: Improvements and Applications," *Science*, pp.413-415, April 26, 1968.

The techniques of radioactive dating have improved since 1950. But greater accuracy reveals discrepancies. The C^{14} dates of cave paintings (which characterize the Upper Paleolithic or late stone age) in Europe seem too recent (11,000 to 16,000 B.C.). Estimated C^{14} dates correspond with historical dates in the first millenium B.C., but not in the second. And the divergence increases with the age of the specimen. Egyptology provides the standard measure. In 1969, at the British Museum, C^{14} counts were made on well-dated specimens from Egypt, including reeds from the Pyramid of Sesostris II (1897–1878 B.C.) and rope from the funerary boat of Cheops (2900 B.C.). The reeds dated a century late; the rope, 450 years late. The basic assumption that C^{14} production in the atmosphere has been constant no longer seems tenable.

Tree rings have long given accurate counts of years and records of climatic conditions. The 1954 discovery in California of living bristlecone pine trees more than 4000 years old suggested a way to establish expected C^{14} counts for a long sequence of well-dated organic materials. Dead samples whose rings overlapped the ring sequence of living trees gave carbon counts extending back 7000 years. The errors cited above were verified and checked with samples of longlived trees in various parts of the world. The Stuiver-Seuss correction curve, based on growth-ring chronology, now allows more reliable radiocarbon dating. The error is less than a century even in the fifth millenium B.C.

Changes in C^{14} production imply variations in the intensity of cosmic radiation and raise questions concerning causal relations among solar activity, sunspots, cosmic rays, C^{14} production, and climate. The Lost City meteorite (see Section 1.7) supplied new kinds of data. Its last orbit around the sun was calculated from the photograph of its meteor stage. As a meteoroid in space, it was exposed to cosmic radiation that produced radioactive isotopes of different half-lives. Disintegration counts after the fall indicate the times at which the isotopes were formed. Times before fall correspond to orbital positions in space. Since the radiation intensity needed to produce specific isotopes from iron and calcium is known, radiation intensity at particular times and distances from the sun may be determined.

SUGGESTED READINGS

Deevey, E. S., Jr., "Radioactive Dating," *Scientific American*, **186**, 2, 24–28, Feb. 1952.

Ferguson, C. W., "Bristlecone Pine; Science and Esthetics," *Science*, 839-846, Feb. 23, 1968.

Fireman, Edward L., "The Lost City Meteorite," *Sky and Telescope*, **39**, 3, 154-158, March 1970.

Hawkes, Jacquetta, "New Dates for Old Times," *The Sunday Times*, London, p. 8, Dec. 14, 1969.

Hurley, P. M., "Radioactivity and Time," *Scientific American*, **181**, 2, 48–51, Aug. 1949.

Libby, Willard F., *Radiocarbon Dating*, 2nd edition. Chicago: University of Chicago Press, 1955.

Putnam, J. L., *Isotopes*. Harmondsworth, England: Penguin Books, 1960.

Symposium on Radioactive Dating, Athens, 1962. Vienna: International Atomic Energy Agency, 1963.

1.13 THE SEPARABLE EQUATION $y' = f(x)\,q(y)$

Both differential equations considered so far are special cases of the first-order equation $y' = f(x, y)$. One other special case lends itself to the same type of theorem. The *separable equation* $y' = f(x)q(y)$ is a natural extension of the previous cases. If $q(y) \neq 0$ on J, the equation may be written

$$y'(x)\,\frac{1}{q(y)} = f(x).$$

Familiarity with integral functions as antiderivatives [see Statement S(1.3.2)] suggests the relation

$$\int_{y_0}^{y(x)} \frac{1}{q(u)}\,du = \int_{x_0}^{x} f(u)\,du.$$

The integral expressions are equal at x_0 if $y(x_0) = y_0$. The reader may show that this relation formally satisfies the separable equation by implicit differentiation with respect to x. But formal differentiation may produce meaningless results; the relation may not define a differentiable function (see Exercise 1 below). This question, whether the relation defines a solution $s: x \to y$ on $I \supset \{x_0\}$, is the burden of the proof of Theorem 1.13.1. The integrals, interpreted as areas, do provide a geometric interpretation of a function so defined. The number x determines an algebraic area (Fig. 1.11a), and the equivalent area in Fig. 1.11(b) determines a

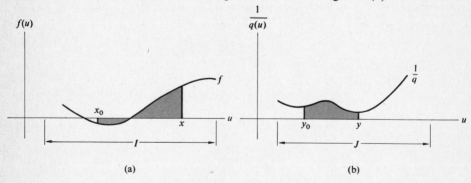

(a)   (b)

Figure 1.11

unique value of y. However, it is possible to sketch functions f on I and q on J such that the integral relation does not define a function on I (see Exercise 2 below). Since the separable equation includes $y' = f(x)$ and $y' = q(y)$ as special cases, the conditions which ensure a solution must be at least as stringent as in the previous theorems.

EXERCISES [1.13.1]

· ·

1. To show that formal differentiation may lead to meaningless results, let

$$f_1: x \rightarrow x, \qquad x \text{ rational and } x \in R$$

$$x \rightarrow -x, \qquad x \text{ irrational and } x \in R,$$

and

$$f_2: x \rightarrow x^2, \qquad x \in R.$$

a) Show that the composite function

$$f_2 \circ f_1(x) = f_2(f_1(x)) = x^2.$$

b) Without considering the functions involved, differentiate the composite function formally to obtain

$$f_1'(x) = \frac{x}{f_1(x)} = \begin{cases} +1 & \text{if } x \text{ rational,} \\ -1 & \text{if } x \text{ irrational.} \end{cases}$$

c) Is f_1 a differentiable function?

d) Pinpoint the mathematical error that led to this confusion.

2. Sketch two functions f_3 and f_4 on intervals I and J such that the relation

$$\int_{y_0}^{y} f_4(u) \, du = \int_{x_0}^{x} f_3(u) \, du$$

does not define a function $f_5: x \rightarrow y$ on $I_0 \supset \{x_0\}$. Justify your answer.

· ·

Theorem 1.13.1 [on solutions of $y' = f(x)q(y)$]

Let f be continuous on the interval I.

Let q be continuous and nonzero on the interval J.

Let y_0 be an interior point of J, and $x_0 \in I$.

Then there exists a unique function with domain $I_0 \subset I$ and range J, satisfying the equation $y' = f(x)q(y)$ and the initial condition $y(x_0) = y_0$.

This solution is defined by the relation

$$\int_{x_0}^{x} f(u) \, du = \int_{y_0}^{y} \frac{1}{q(u)} \, du, \qquad x \in I_0. \tag{1.13.1}$$

Proof: Assume that $q(y) > 0$ for $y \in J$. Then the relation (1.13.1) defines a function $s: x \to y$ on $I_0 \subset I$ with range J. Rephrased, the mapping

$$x \to \int_{x_0}^{x} f(u) \, du = \int_{y_0}^{y} \frac{1}{q(u)} \, du \to y$$

is well defined. Given any $x \in I$, the number $\int_{x_0}^{x} f(u) \, du$ is uniquely defined, since f is continuous on I. Indeed, the correspondence defines a differentiable function, call it f_1 on I (see the function diagram, Fig. 1.12). For any $y \in J$, the number

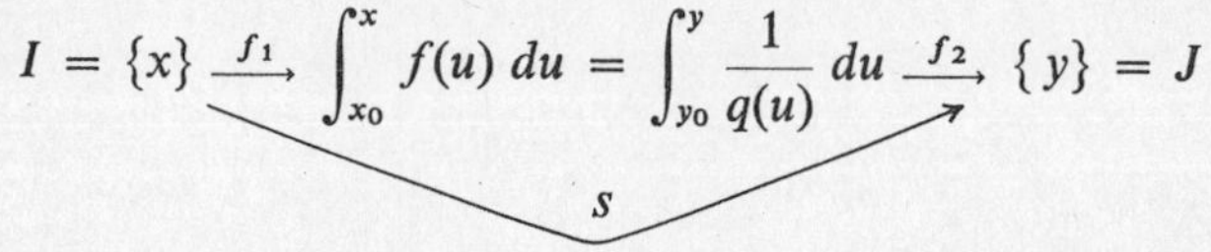

$$I = \{x\} \xrightarrow{\ f_1\ } \int_{x_0}^{x} f(u) \, du = \int_{y_0}^{y} \frac{1}{q(u)} \, du \xrightarrow{\ f_2\ } \{y\} = J$$

$$s$$

Figure 1.12

$\int_{y_0}^{y} 1/q(u) \, du$ is uniquely defined, and, since the integrand is positive, the correspondence defines an increasing differentiable function on J. Hence the inverse function, call it f_2, exists and is itself increasing and differentiable. Thus to each number $\int_{y_0}^{y} 1/q(u) \, du$ in the domain of f_2 there corresponds exactly one number y. In particular

$$x_0 \xrightarrow{\ f_1\ } \int_{x_0}^{x_0} f(u) \, du = 0 = \int_{y_0}^{y_0} \frac{1}{q(u)} \, du \xrightarrow{\ f_2\ } y_0.$$

If y_0 is an *interior* point of J (not an endpoint), then for some ε, the interval $(y_0 - \varepsilon, y_0 + \varepsilon) \subset J$. And the interval

$$(\alpha_1, \beta_1) = \left(\int_{y_0}^{y_0 - \varepsilon} \frac{1}{q(u)} \, du, \ \int_{y_0}^{y_0 + \varepsilon} \frac{1}{q(u)} \, du \right)$$

contains 0 and is in the domain of f_2. Since f_1 is continuous, and $f_1(x_0) = 0$, there is an interval I_0, in I and containing x_0, such that $x \in I_0$ implies that

$$f_1(x) = \int_{x_0}^{x} f(u) \, du \in (\alpha_1, \beta_1).$$

Thus on I_0, the function

$$s(x) = f_2(f_1(x)) = y$$

is well defined. As the composite of two differentiable functions, it is differentiable and satisfies the differential equation as well as the initial condition.

To show the uniqueness,† suppose there is a function w on I_0 such that

$$w'(x) = f(x)\, q(w(x)) \quad \text{and} \quad w(x)_0 = y_0.$$

If w and s are distinct, then for some $x_\alpha \in I_0$,

$$y(x_\alpha) \neq w(x_\alpha).$$

And since $1/q$ is positive,

$$\int_{y_0}^{y(x_\alpha)} \frac{1}{q(u)}\, du \neq \int_{y_0}^{w(x_\alpha)} \frac{1}{q(u)}\, du.$$

But the functions

$$p : x \to \int_{y_0}^{y(x)} \frac{1}{q(u)}\, du \quad \text{and} \quad p_w : x \to \int_{y_0}^{w(x)} \frac{1}{q(u)}\, du,$$

both on I_0, have the same derivative function because

$$p'(x) = D_x \int_{y_0}^{y(x)} \frac{1}{q(u)}\, du = \left[\frac{1}{q(y)} \right] y'(x)$$

$$= \frac{1}{q(y)} f(x)\, q(y) = f(x),$$

and similarly for $p_w'(x)$.   Hence $p(x)$ and $p_w(x)$ differ by a constant.   But $p(x_0) = p_w(x_0)$. Therefore $p(x) = p_w(x)$ for $x \in I_0$. In particular, $p(x_\alpha) = p_w(x_\alpha)$. That is, $w(x) = s(x) = y(x)$ on I_0.
    The proof for $q(y) < 0$ is parallel.   ■

EXERCISES [1.13.2]

· ·

For each of the differential equations listed in the Exercises 1 through 6,
   a) Sketch the direction field.
   b) Determine maximum intervals I and J for which the hypothesis of Theorem 1.13.1 is satisfied.
   c) Determine solutions through the given points and indicate domains of the solution functions. Can a largest domain be determined in each case?

---

†Note that this is not as simple as the previous situations. In Theorems 1.3.1 and 1.8.1, two different solution functions (or their inverses) had to be antiderivatives of the same fixed function. Here, two functions may be distinct and have distinct derivatives and yet each may satisfy conditions of the type given. For example, on the domain $x \geq 0$, both the functions $y_1(x) = 0$ and $y_2(x) = x^3$ satisfy the conditions $y' = 3xy^{1/3}$ and $y(0) = 0$.

*d) Indicate what happens when I and J are enlarged so that the hypothesis of Theorem 1.13.1 is no longer satisfied. Are there several solutions through a single point? Is there a solution through every point? Are extended solution curves still differentiable? Can solutions still be defined by the relation

$$\int_{x_0}^x f(u)\ du = \int_{y_0}^y \frac{1}{q(u)}\ du?$$

1. $y' = xy$, $(0, 1)$ and $(0, 0)$.

2. $y' = -x/y$, $(0, 1)$ and $(0, 0)$.

3. $y' = xy^{1/2}$, $(1, 1/16)$ and $(0, 0)$.

4. $y' = y/x$, $(1, 1)$ and $(1, 0)$.

5. $y' = y \sin x$, $(\pi/2, 1)$ and $(\pi, 1)$.

6. $y' = (\sin x)/y$, $(\pi/2, 1)$ and $(0, 1)$.

7. Consider the equation $y' = y \sin x$. Show that if $I = R$ and $J = [1, 2]$, there does not exist a solution through the point $(\pi, 1)$, but there is a solution through $(\pi, 2)$. Discuss the condition that y_0 be an interior point. Is it sufficient and/or necessary in Theorem 1.13.1?

. . . .

1.14 ESCAPE VELOCITY

Theorem 1.13.1 allows the consideration of another aspect of motion in the earth's gravitational field: the escape velocity. What is the initial velocity necessary to send a satellite into orbit, a rocket to the moon, or an object into space? Over such large distances, the gravitational acceleration may not be considered as constant.

It has been verified experimentally that the gravitational attraction between two bodies varies directly as the product of their masses (m) and inversely as the square of the separation distance (r). Each body is pulled toward the other with a force of magnitude

$$|F| = \frac{km_a m_b}{r^2}.$$

A derivation of this inverse-square law, based on planetary motion, will be given in the next section. Only when the total motion is such that $1/r^2$ remains constant, within the required degree of accuracy, is the assumption of a constant gravitational acceleration realistic.

In the mathematical model, each body is replaced by a point indicating the position of the centroid. To consider the motion of an object relative to the earth, the point E representing the earth's position, is taken as the origin of the coordinate system (Fig. 1.13). For simplicity, assume that the initial velocity of the object, at position P, has the same direction as the ray $\overrightarrow{EP}$. The gravitational force F, of the earth on the object, has direction $\overrightarrow{PE}$. In this third model, no other forces are

Figure 1.13

considered. The motion starts and continues on the line $\overleftrightarrow{EP}$. Hence a one-dimensional system with coordinate r, denoting the distance from E to P, is sufficient. If the direction $\overrightarrow{EP}$ is taken as positive, then

$$F = -\frac{km_E m_P}{r^2}.$$

By Newton's law, the acceleration of the object is

$$a = -\frac{km_E}{r^2}.$$

The constant of proportionality, k, can be evaluated since the acceleration, g, at the surface of the earth, $r = r_E$, is known: $-km_E = gr_E^2$. As before, if the functions defined by the position $r(t)$ and velocity $v(t)$ are assumed to be differentiable, then by definition $\dot{v} = a$ and $\dot{r} = v$. Thus the motion satisfies the condition

$$a = \dot{v} = gr_E^2(1/r^2). \tag{1.14.1}$$

In this form, the differential equation does not fit any of the theorems considered so far: $\dot{v}$ is not expressed as a function of t and/or v. Three variables, t, r, and v are involved. The original question asks for the initial velocity needed to send an object a given distance, and thereby defines velocity and distance as the quantities of interest. Hence the most useful solution would be a velocity function $v = v(r)$. Can the usual velocity function $v_f: t \to v$ be replaced by a velocity function $\hat{v}: r \to v$ on $I_r = [r_E, r_c]$, where r_c is the desired final distance? The problem of sending an object away from the earth restricts interest to the region of positive velocity. In this region $r_f: t \to r$ is an increasing function. Hence the inverse $r_f^{\leftarrow}: r \to t$ is well defined and differentiable. The composite function $v(r^{\leftarrow}(r)) = v$ defines a function $\hat{v}: r \to v$ on $[r_E, r_c] = r_f[0, t_c]$ (see Fig. 1.14). Note that $v(t) = v(r)$ where $r = r(t)$.

$$I_r = \{r\} \xrightarrow{\;r_f^{\leftarrow}\;} T = \{t\} \xrightarrow{\;v_f\;} V = \{v\}$$
$$\hat{v}$$

Figure 1.14

The function values are the same for corresponding values of t and r; the sets of number pairs $\{(t, v)\}$ and $\{(r, v)\}$, representing v_f and $\hat{v}$, are different. Here the chain rule in the notation

$$\frac{dv}{dt} = \frac{dv}{dr}\frac{dr}{dt}$$

is particularly suggestive: $\hat{v} = v'(r)\,v$. Hence equation (1.14.1) may be written

$$v'(r) = gr_E^2 \left(\frac{1}{r^2}\right)\frac{1}{v},$$

and this is a separable differential equation.

EXERCISES [1.14.1]

1. Find the initial velocities necessary to send an object
   a) to the moon,
   b) into an orbit 100 km above the earth's surface,
   c) into space. The region effectively outside the earth's gravitational field is represented by letting the distance $r \to \infty$.

   [*Suggestions*: Let $v_0 = v(r_E)$ be the positive velocity imparted by the launching at the earth's surface. The "necessary initial velocity" implies a minimal velocity: i.e., the final velocity at the desired distance may be zero. The most convenient form of the solution may be to express v_0 as a function of v and r and to consider the limit as $v \to 0$. Use approximations that are reasonable but allow simple computations.]

2. Do the velocities in Exercise 1 imply a significant difference in the techniques of launching?

*3. Sketch the function curve of $\hat{v}: r \to v$ for $v_0 = 10, 11, 12$ km/sec.

*4. Consider the same equation in the region $v < 0$. Ignoring air resistance, show that the particle velocity at r_E of an object dropped at a distance r is $v = \sqrt{2r_E|g|}\,(1 - r_E/r)^{1/2}$.

If nonconstant gravitational acceleration and air resistance are both considered, the problem is more complicated. The drag changes as the air density changes, and there is no simple functional representation of the variation. Solutions are obtained as numerical approximations. Numerical methods allow the introduction of appropriate drag values based on empirical evidence.

1.15 THE LAW OF GRAVITATION

Each of the three models for motion in a gravitational field is based on an assumption concerning the gravitational force. Once the quantitive measures of the forces are known, Newton's law yields a differential equation. Successive solutions provide quantitative expressions for velocity and position, and thus the motion is determined.

The actual derivation of the quantitative law of gravitation, one of Isaac Newton's great accomplishments, illustrates the reverse procedure. The gravitational force is not observable as an ordinary push or pull, but the resulting motion is observable. In the seventeenth century, the motion of the planets was known; the problem was to determine the forces acting. Tycho Brahe (1546–1601) had devised instruments and recorded precise measurements of the planets' positions over a period of 20 years. His timepiece was a clepsydra. Johannes Kepler (1571–1630) studied these observations for some 16 years and deduced three laws or generalizations concerning planetary motion. Thus planetary motions were known through direct observations and were understood in terms of simple laws. Two of these laws are sufficient for this derivation.* It was also recognized that the earth exerts a pull on objects near its surface, and that this same pull keeps the moon in its orbit. The general problem of determining these gravitational forces was well stated by Newton.†

If this force was too small, it would not sufficiently turn the moon out of a rectilinear course; if it was too great, it would turn it too much, and draw down the moon from its orbit towards the earth. It is necessary, that the force be of a just quantity, and it belongs to the mathematicians to find the force, that may serve exactly to retain a body in a given orbit, with a given velocity; and vice versa, to determine the curvilinear way, into which a body projected from a given place, with a given velocity, may be made to deviate from its natural rectilinear way, by means of a given force.

This question occupied other well-known men of the period, including Christopher Wren, Christian Huygens, Robert Hooke, and Edmund Halley.

---

*Kepler's first two laws: 1. The planetary orbits are ellipses with the sun as a focus. 2. As a planet moves about the sun, an imaginary line segment from the sun to the planet sweeps out equal areas in equal times.

†Sir Isaac Newton, *Mathematical Principles*, Definition V, Cajori's revision of Motte's English Translation (1729), page 4, Berkeley, Calif.: University of California Press, 1934.

Cambridge University closed during the years of the great plague, 1664–1665. Newton, then in his early twenties, spent these years in the isolation of his mother's farm. Here he laid the groundwork for the solution of this particular problem. In particular, he stated clearly the assumptions necessary to set up an effective mathematical model. In addition to setting up the mechanics, that is, the relations between forces and motion, he created the calculus to serve as a model of the non-uniform motion exhibited by the planets.

Newton's work was written in Latin and in the framework of Euclidean geometry, the then accepted languages of science and mathematics. The derivation considered here is parallel in that the inverse-square law of gravitation is deduced from two of Kepler's laws; but the framework is that of today's calculus. The reverse implication, that the inverse-square law of gravitation implies Kepler's first two laws, will be given in the chapter on linear equations.

For the mathematical model, let the sun be represented as the point S and the planet as point P. Since the elliptical path of motion and its focus S are in the same plane, a two-dimensional coordinate system suffices. Let the origin be at S. Each point P representing a position of the planet may be indicated by coordinates (x, y) or (r, θ). The polar-coordinate system is the more natural framework for Kepler's second law. Moreover, the final result, that is, the magnitude of the gravitational force, is expected to be a function of the separation distance r. If the position vector **P** is a smooth function of time, that is, the components x and y, or r and θ, are twice-differentiable functions of t, then the acceleration vector **a** at any point P is $(\ddot{x}, \ddot{y})$ and the components may be expressed as functions of r and θ and their derivatives.

1. *Kepler's second law implies the differential equation* $r^2\dot{\theta} = k$.

Consider alternative statements of Kepler's law. In equal time intervals, the radius vector sweeps out equal areas, regardless of the initial position of P (see Fig. 1.15).

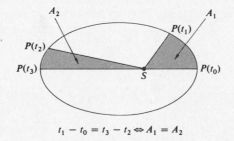

$$t_1 - t_0 = t_3 - t_2 \Leftrightarrow A_1 = A_2$$

Figure 1.15

Or, for a given change in time, the change in area is constant. If the area $A = A(t)$ is a differentiable function of time, then $\dot{A} = K$. In this instance, it is desirable to express $\dot{A}(t)$ as a function of r and θ and their derivatives.

The reader may be familiar with the integral expression for area in polar coordinates,

$$A(\theta) = \int_{\theta_1}^{\theta} \tfrac{1}{2}r^2(u)\, du,$$

where A is the area in the region bounded by the curve $r = f(\theta)$ and the radius vectors determined by θ_1 and θ. If so, then it follows from the fundamental theorem, Statement S(1.3.2), that $\dot{A} = D_\theta A\, D_t \theta = \tfrac{1}{2}r^2(\theta)\, \dot\theta$. Hence, if $\dot{A}$ is constant, $r^2\dot\theta = k$. (For an elementary derivation of $\dot{A} = \tfrac{1}{2}r^2\dot\theta$, see Appendix A.5.)

2. Kepler's second law implies a central force.

The equation $r^2\dot\theta = k$ may be expressed in terms of the coordinates x, y, and their derivatives by using the basic relations between rectangular and polar coordinates. The equivalent equation is $\dot{y}x - \dot{x}y = k$. Differentiating again yields $\ddot{y}x = \ddot{x}y$. Hence the components of the position and acceleration vectors are proportional, and $\mathbf{a} = (\ddot{x}, \ddot{y}) = k(x, y) = k\mathbf{P}$. That is, at every position P, the acceleration (and hence the force) is directed along the radius vector on the fixed point S. In such a case, the body is said to be subject to a *central force*. And since the path of motion curves away from the tangent on the side toward S, the force must be directed toward S. This suggests that the motion is the result of a pull exerted by the sun.

3. Kepler's second law implies that $|\mathbf{a}| = |\ddot{r} - k^2/r^3|$.

The acceleration vector $\mathbf{a} = (\ddot{x}, \ddot{y})$ may be expressed in terms of r and θ and their derivatives. Express x and y in terms of r and θ, and differentiate, using the product rule and the chain rule, or the chain rule for a function of two variables. At each step, $\dot\theta$ may be replaced by k/r^2 (using the differential equation $r^2\dot\theta = k$). Thus the reader may ascertain that

$$(\ddot{x}, \ddot{y}) = [\ddot{r} - k^2/r^3]\,(\cos\theta, \sin\theta)$$

and hence

$$|\mathbf{a}| = |\ddot{r} - k^2/r^3|.$$

4. Kepler's first law $\Rightarrow$ the equation $\delta/r = 1/e + \cos\theta$.

An ellipse may be defined in terms of a fixed point and a fixed line (not on the point), called the focus and the directrix. The ellipse is the set of points $\{P(r, \theta)\}$ such that the ratio of the distance from P to the focus and the distance from P to the directrix is a constant $e < 1$. The eccentricity e determines the shape of the ellipse. Let the path of the planet be defined by the focus S, directrix $x = \delta$ (see Fig. 1.16) and eccentricity e. This ellipse is arbitrary except that its major axis is on

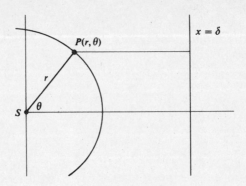

Figure 1.16

the horizontal axis in order to simplify the analytic expressions. Then, by definition, for any point $P(r, \theta)$ on the ellipse, $r \neq 0$ and

$$\frac{r}{\delta - r \cos \theta} = e \quad \text{or} \quad r = \frac{e\delta}{1 + e \cos \theta}.$$

For differentiation, the form $\delta/r = 1/e + \cos \theta$ is simpler.

5. *Kepler's first and second laws imply that* $|\mathbf{a}| = k_1/r^2$.

In order to express $|\mathbf{a}|$ as a function of r, it is only necessary to express $\ddot{r}$ in terms of r. Since r and θ are coordinates of a point on the ellipse, $\delta/r = \cos \theta + 1/e$ (and $r \neq 0$), $\ddot{r}$ may be obtained by differentiating this equation with respect to t, and replacing each $\dot{\theta}$ by k/r^2. Proceeding in this manner, the reader may show that $|\mathbf{a}| = (k^2/\delta r^2)(\cos \theta - \delta/r)$. Finally, applying the equation of the ellipse,

$$|\mathbf{a}| = -\frac{k^2}{e\delta \, r^2} = \frac{k_1}{r^2}.$$

Thus, if the motion of one body about another satisfies Kepler's first two laws, the magnitude of the acceleration (and hence the force) varies inversely as the square of the separation distance.

SUGGESTED READINGS

Andrade, E. N. da C., *Sir Isaac Newton*, Garden City, N.Y.: Doubleday, 1958.

Holton, Gerald, *Introduction to Concepts and Theories of Physical Science*, Chapter 11. Reading, Mass.: Addison-Wesley, 1952.

Koestler, Arthur, *The Sleepwalkers*. London: Hutchinson, 1959.

1.16 OTHER FIRST-ORDER EQUATIONS

The three theorems considered so far are special in several respects. The third theorem includes the first two as special cases, so it suffices to consider Theorem 1.13.1. This theorem states the conditions under which a separable differential equation has a solution which satisfies a given initial condition, and is unique. In addition, it is a very practical theorem. It tells exactly how the solution function is defined. The analogous theorem for the more general first-order equation $y'(x) = f(x, y)$ concerns only the existence and uniqueness of solutions. A proof involving sequences and series is given in Chapter 4; it also provides a method of numerical approximation of limited application. The theorem says nothing about how to find solutions in the general case. Methods of solution depend on the form of $f(x, y)$; ingenious methods have been devised for particular types of equations. Books on elementary differential equations used to consist of a compilation of such special techniques. Such methods are interesting to one who is already seeking solutions for differential equations. Otherwise, these specific techniques can be looked up when needed—if the basic ideas concerning existence, uniqueness, and effective domains of solution functions are already understood. Here, only two of these techniques are illustrated.

The first technique, applicable to homogeneous differential equations, is another illustration of how an equation may be reduced to a separable equation, that is, to a previously solved problem, by substitution of another function (see Section 1.4). The particular homogeneous equation considered here is a model for a reflector which directs or collects light or radio waves. Today, with the techniques of radio astronomy, such reflectors permit the study of motions previously obscured to us, such as the motion of the spiral arms of our own galaxy.

The second technique, that of integrating factors, is applied to first-order linear equations. Although it is often a "hunt and find" technique, it has proved to be of general use and is worth adding to one's bag of tricks.

*1.17 THE REFLECTOR

Directed light beams, radar antennas, and radio telescopes are all based on the principle of concentrating energy so that it may be detected at or from great distances. In an automobile headlight, the reflecting surface is shaped to reflect the incident light from the bulb in a single direction. A radio telescope, on the other hand, collects radio signals from a single direction and reflects them to a single point. Thus the signal is strengthened so that it may be received and interpreted. The radar antenna is used in both ways: it transmits directed waves and receives those that are reflected back from outside objects. How must these reflectors be shaped?

Certain basic laws of reflection, determined by experiment, are necessary for our model.

1. The incident ray, the reflected ray, and the normal to the reflecting surface at the point of incidence all lie in the same plane.

2. For a curved surface, reflection is effectively determined by the tangent plane at the point of incidence.

3. The measures of the angles of incidence and reflection, each determined with respect to the normal, are equal.

   The problem itself, and the laws above, imply a surface of great symmetry. In the model, a point S represents the source of energy, and the rays reflected from the reflecting surface are assumed to be parallel (and in the same direction). Let L be a line on S having the required direction. Let Π_L be any plane on L. Let C be the intersection of the surface and Π_L (see Fig. 1.17). Since, for any point P on C, Π_L

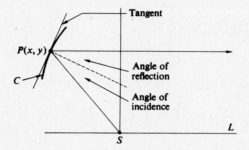

Figure 1.17

contains the incident ray and reflected ray, it also contains the normal to the surface at P (law 1, above). Therefore, the angles of incidence and reflection lie in this plane. Thus, in the plane Π_L, it is possible to determine the tangent to the curve C, at any point, so that the reflected ray is parallel to L. For any other plane Π_{L_α} on L, the situation is identical. Hence the same curve C may be expected in each Π_L. Therefore the reflecting surface must be a surface of revolution: a two-dimensional mathematical model, to determine C in a single plane, will suffice.

   Let S be the origin and L the horizontal axis. Assume that the curve C is smooth, that is, C is the curve of a differentiable function. Then at each point $P(x, y)$ on C there is a tangent to C, having slope $y'(x)$.

EXERCISES [1.17.1]

. . . .

1. Set up a differential equation by imposing the third law of reflection as a condition on the solution function $s: x \to y$ representing the curve C. [*Hint:* It may be easier to deal with tangents of angles rather than the angles themselves.]

. . . .

In order to express the differential equation (see Exercise 1 above)

$$y'^2 + 2 \left(\frac{x}{y}\right) y' - 1 = 0$$

in the form $y' = f(x, y)$, it may be considered as a quadratic equation in y' with solutions

$$y' = -\frac{x}{y} \pm \left[\left(\frac{x}{y}\right)^2 + 1\right]^{1/2}, \qquad (1.17.1)$$

or as the product of two differential equations

$$\left[y' + \frac{x}{y} - \left(\frac{x^2}{y^2} + 1\right)^{1/2}\right]\left[y' + \frac{x}{y} + \left(\frac{x^2}{y^2} + 1\right)^{1/2}\right] = 0.$$

A solution function that makes one of the factors zero, makes the product zero and hence satisfies the equation

$$y'^2 + 2(x/y) y' - 1 = 0.$$

Slopes and tangents do not impose a direction; $\tan \theta = \tan \theta + \pi$. Another condition is needed to ensure a directed beam. The incident and reflected rays must be on the same side of the surface as in Fig. 1.18. Hence for $y > 0$, the condition

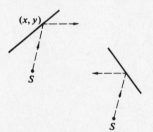

Figure 1.18

$y' > 0$ ensures that the horizontal ray is reflected to the right. Since

$$\left[1 + \left(\frac{x}{y}\right)^2\right]^{1/2} > \left|\left(\frac{x}{y}\right)\right|,$$

in each differential equation in (1.17.1), y' is always positive (or negative). Hence the choice of a single equation (with $+$ or $-$ sign) in effect selects a solution (in the upper half-plane) which directs the beam to the right or the left. Physically, a horizontal surface cannot produce a horizontal reflected ray; hence the restriction $y' \neq 0$ is reasonable.

*1.18 THE HOMOGENEOUS DIFFERENTIAL EQUATION

Neither first-order equation in (1.17.1) is separable, but each is equivalent to a separable equation because of the property that $y' = f(y/x)$. An equation which can be expressed in this form is called a *homogeneous differential equation*.* Consider the function $v(x) = y/x$, $y > 0$ (or $y < 0$). Then

$$y(x) = xv(x) \quad \text{and} \quad y' = v'(x)x + v.$$

But $y' = f(v)$, hence

$$v'(x) = \frac{f(v) - v}{x},$$

which is separable. And once $v(x)$ is determined, $y(x)$ is well defined on the same domain.

EXERCISES [1.18.1]

. . . .

1. Show that for (x_0, v_0) in the region $\{(x, v): x < 0, v < 0\}$ the solution of the separable equation equivalent to

$$y' = -\frac{x}{y} + \left[\left(\frac{x}{y} \right)^2 + 1 \right]^{1/2}$$

is defined by the relation

$$\ln \left(\frac{1 + \sqrt{v_0^2 + 1}}{1 + \sqrt{v^2 + 1}} \right) = \ln \frac{x}{x_0}.$$

2. Obtain the solution $s: x \to y$ through (x_0, y_0) in the region $\{(x, y): x < 0 \text{ and } y > 0\}$.

3. Show that the curve C, defined by s is a piece of a parabola. (Replace complicated constants by K and write in standard form.) What relation does the point S (Fig. 1.17) have to the parabola?

4. Find the points $(0, y)$ and $(x, 0)$ on C by using the limit convention; i.e., determine

$$\lim_{x \to -0} y \quad \text{and} \quad \lim_{y \to +0} x.$$

5. Determine the solution through $(-1, \sqrt{3})$. Sketch the curve.

6. Find a solution in the region $\{(x, y): x > 0, y > 0\}$ which, together with the solution in Exercise 5 and its limit points, forms a differentiable function in the region $\{(x, y): y > 0\}$ satisfying the differential equation and the initial condition $y(-1) = \sqrt{3}$.

---

*The usual definition, $f(x, y)$ is *homogeneous of degree n* if $f(kx, ky) = k^n f(x, y)$, offers a useful test as well. The word "homogeneous" has various meanings in mathematics. A second meaning is used in Chapter 2.

[In Exercise 6, the conditions $x \neq 0$ and $y \neq 0$ merely reflect the limitations of a system which does not assign a number to a vertical slope.]

7. Find a solution in the region $[(x, y): x < 0, y < 0]$ which, together with the solution of Exercise 6, makes a smooth curve. [The curve may not represent a function.]

8. Determine and sketch the solution curve C of the equation

$$y' = -\frac{x}{y} - \left[\left(\frac{x}{y}\right)^2 + 1 \right]^{1/2}$$

for the initial condition $y(-1) = \sqrt{3}$.

9. Describe the shape that a three-dimensional reflector must have in order to produce a directed beam.

. . . .

The radio-telescope dish at Jodrell Bank, England, is about 75 m in diameter. With such large dishes, the 21-cm waves emitted by the excited hydrogen atoms in the spiral arms of our galaxy can be detected. The Doppler shift in wavelength is used to determine the velocities of these hydrogen clouds.*

1.19 INTEGRATING FACTORS FOR FIRST-ORDER LINEAR DIFFERENTIAL EQUATIONS

The equation $y' + p(x)y = r(x)$, where p and r are specified functions on an interval I, is called a *first-order linear differential equation*, since it is of first degree in y and y'. Some of the separable equations already solved are also linear. Specifically, consider the equations and solutions listed below.

Equation	Initial condition	Solution	Exercise or equation
$y' - y = 0$	$y(0) = 1$	$y = e^x,\ x \in R$	2[1.2.1]
$y' - xy = 0$	$y(0) = 1$	$y = e^{x^2}/2,\ x \in R$	1[1.13.2]
$y' - (1/x)y = 0$	$y(1) = 1$	$y = x[\ = e^{\ln x}],\ x \in R^+$	4[1.13.2]
$y' - (\sin x)y = 0$	$y(\pi/2) = 1$	$y = e^{-\cos x},\ x \in R$	5[1.13.2]
$\dot{v} + (k/m)v = g$	$v(0) = 0$	$v = K_1 - K_1 e^{-(k/m)t},\ t \in [0, t_c]$	(1.7.1)

The common element in their solutions suggests a general method of obtaining solutions.

---

*See Abell, *Exploration of the Universe*, 2nd ed., pages 530-536. New York: Holt, Rinehart, and Winston, 1969.

In each case, the solution contains the term $\exp\left[-\int_{x_0}^{x} p(u)\, du\right]$, where $\exp u = e^u$. This term, introduced as a factor in the differential equation, has a special effect on the binomial $y' + p(x)y$. Specifically,

$$y' \exp\left[\int_{x_0}^{x} p(u)\, du\right] + p(x)y \exp\left[\int_{x_0}^{x} p(u)\, du\right]$$

is the derivative of the product function

$$w(x) = y(x) \exp\left[\int_{x_0}^{x} p(u)\, du\right].$$

If $r(x) = 0$, then $w'(x) = 0$, and hence by the mean-value theorem, $w(x) = K$. And the initial condition $y(x_0) = y_0$ determines the constant K as y_0. Thus the solution is specified as

$$y = y_0 \exp\left[-\int_{x_0}^{x} p(u)\, du\right].$$

The general first-order linear equation $y' + p(x)y = r(x)$, after introduction of the *integrating factor* $\exp\left[\int_{x_0}^{x} p(u)\, du\right]$, becomes

$$w'(x) = r(x) \exp\left[\int_{x_0}^{x} p(u)\, du\right],$$

and Theorem 1.3.1 is applicable.

EXERCISES [1.19.1]

. . . .

Find an integrating factor for each of the following equations and obtain a solution through the given point.

1. $y' + (1/x)y = e^x$, (x_0, y_0)
2. $(1 + x^2)y' + xy = (1 + x^2)^{1/2}$, $(0, 1)$
3. $y' + (\cos x)y = x$, $(\pi/2, 0)$
4. Show that the solution function satisfying the general first-order linear equation $y' + p(x)y = r(x)$ and the initial condition $y(x_0) = y_0$ may be specified as

$$y = \exp\left[-\int_{x_0}^{x} p(u)\, du\right]\left[\int_{x_0}^{x} r(u) \exp\left[\int_{x_0}^{u} p(v)\, dv\right] du + y_0\right].$$

(In practice it is easier to obtain the integrating factor for the individual equation and solve the equation $w'(x) = r(x) \exp\int_{x_0}^{x} p(u)\, du$. The general form of the solution function above is too cumbersome.)

. . . .

Integrating factors are used in various types of differential equations. The form of the factor in each case may differ. Often the integrating factor is chosen by inspection. Specific integrating factors have been developed for certain classes of equations, such as the first-order linear equations.

Linear equations of order n are central to the study of differential equations. Second-order linear equations are the main concern of the next chapter.

CHAPTER 2

SECOND-ORDER LINEAR EQUATIONS

2.1 INTRODUCTION

Linear equations comprise a "comfortable" class of differential equations. A relatively complete theory has been developed. Existence and uniqueness questions have answers, and the theory provides general and practical methods of obtaining solutions. It is comfortable to ask questions with ready answers. Nonlinear equations are more difficult. There are not many answers, and no corresponding chapter in this book.

However, linear equations offer more than ready solutions. They serve as models for some of the basic ideas in physical science such as wave phenomena and oscillatory motion. They serve as local approximations to nonlinear equations. Linear equations and numerical methods probably offer, so far, the best approach to nonlinear equations. Second-order linear equations illustrate all these aspects; moreover, the basic theory for order two generalizes easily to higher-order linear equations. Hence, for simplicity, the discussion will be limited to second-order equations. But the development will be that which facilitates generalization to order n; extensions to higher orders will be suggested in the exercises.

The general second-order linear differential equation has the form

$$y''(x) + p(x)\, y'(x) + q(x)\, y(x) = r(x),$$

where p, q, and r are specified functions on some interval I. Simple examples are

$$y'' + y = 0, \tag{2.1.1}$$

$$y'' - y' = 0. \tag{2.1.2}$$

In both cases, p, q, r are constant functions on R. Solutions on R can be found by inspection, that is, by recalling functions with the property that $y'' = -y$ or $y'' = y'$. Both the sine and cosine functions satisfy (2.1.1) while any constant function and the exponential function satisfy (2.1.2). Even if the solution function is restricted to include a specific number pair, or point, (x_0, y_0), there are still numerous solutions. For example, all the solutions listed in Table 2.1 contain $(0, 0)$.

Can you add three different solutions through $(0, 0)$?

The basic questions are the same as before. Under what conditions does a solution exist? When is the solution unique? Can all the solutions be written in a

TABLE 2.1

Solutions of (2.1.1)	Solutions of (2.1.2)
$y = \sin x$	$y = e^x - 1$
$y = 2 \sin x$	$y = e^{x+3} - e^3$
$y = \sin(x + \pi/6) - \frac{1}{2} \cos x$	

given form? But clearly, since there are several solutions to (2.1.1) through the same point, the answers will be different.

The existence theorem is surprisingly simple. It requires only the continuity of the given functions. And the solution is unique if, in addition to the fixed point (x_0, y_0), a fixed slope $y'(x_0)$ is required. Hence the existence and uniqueness theorem seems but a slight generalization of Theorem 1.3.1. Yet the class of equations includes many more variations.

Theorem 2.1.1 [on solutions of $y'' + p(x)y' + q(x)y = r(x)$]

Let p, q, r be continuous functions on I.
Let $x_0 \in I$; $y_0, y'_0 \in R$.

Then there is a unique function on I such that

$$y(x_0) = y_0,$$
$$y'(x_0) = y'_0,$$

and $y(x)$ satisfies the equation

$$y'' + p(x)y' + q(x)y = r(x) \qquad \text{for} \quad x \in I.$$

Note that although the conditions for existence and uniqueness are very simple, nothing is said about the form of the solution or a method of obtaining the solution. Hence the gain in one direction is balanced by a loss.

The proof of Theorem 2.1.1 involves sequences and series and will not be given until Chapter 4. But the implications of the theorem are illustrated by equations (2.1.1) and (2.1.2). In both cases, the functions p, q, and r are constant and hence continuous on R. All the solutions listed in Table 2.1 are on R, but have different slopes at $(0, 0)$. According to the theorem, each of these functions is the unique solution having that slope at the point $(0, 0)$.

EXERCISES [2.1.1]

. . . .

For each equation in Exercises 1 through 8, determine by inspection or guesswork solutions

a) through $(0, 0)$,
b) having slope 0 at $(0, 0)$,

c) having slope 1 at $(0, 0)$,
d) having slope 0 at $(1, 0)$.

1. $y'' + 4y = 0$ 5. $y'' + 4y = 1$
2. $y'' + 2y = 0$ 6. $y'' - y' = 2$
3. $y'' - y = 0$ 7. $y'' + 4y = x$
4. $y'' - 4y = 0$ 8. $y'' + y' = x$

. . . .

The following corollary of Theorem 2.1.1 will be useful.

Corollary 2.1.1

Let $r = 0$, and p and q be continuous on I.
Let $x_0 \in I$, $y_0 = y_0' = 0$.

Then the trivial solution

$$y(x) = 0 \quad \text{on} \quad I$$

is the unique solution satisfying

$$y'' + p(x)y' + q(x)y = 0,$$

$$y(x_0) = 0, \quad \text{and} \quad y'(x_0) = 0.$$

In particular, Corollary 2.1.1 explains part (b) of Exercises [2.1.1].

For the present, Theorem 2.1.1 will be assumed, and methods of obtaining solutions will be considered. The solution methods depend on the coefficient functions p, q, and r. Methods for constant coefficients will be developed in this chapter; nonconstant coefficients will be discussed in Chapter 3. However, the very linearity of the equation suggests relations among solutions and an orderly approach to obtaining solutions. These general implications, to be considered first, are best seen through the concept of an operator.

2.2 A LINEAR OPERATOR

An *operator* $\mathcal{O}$ is a mapping from a set of real functions into a set of real functions. In other words, an operator is a function whose domain and range are both sets of functions.

Examples are:

$$\mathcal{O}_1 : f \to f + 2$$
$$\mathcal{O}_2 : f \to \tfrac{1}{2} f$$
$$\mathcal{O}_3 : f \to f' = Df$$
$$\mathcal{O}_4 : f \to \text{the antiderivative function } A(x) = \int_{\frac{1}{2}}^{x} f(t) \, dt.$$

$\mathcal{O}_1$ and $\mathcal{O}_2$ map all real functions into the set of all real functions. $\mathcal{O}_3$, called the differential operator and denoted by D, has for its domain the set of differentiable functions. A domain for $\mathcal{O}_4$ is the set of continuous functions on an interval containing $\frac{1}{2}$; a larger domain is the set of integrable functions on intervals containing $\frac{1}{2}$. Since an operator is a special type of function, addition, multiplication, and composition of operators are defined in the usual way.

The operator which takes twice differentiable functions into their second derivatives is written D^2. This agrees with the notation of differentiation, but it must be remembered that it indicates the composition, not the product of operators, that is,

$$D^2 f = D(Df) = D \circ Df.$$

Similarly

$$D^n : f \to f^{(n)},$$

where $f^{(n)}$ is the nth derivative of f.

In particular, a *linear operator* has the property that

$$\mathcal{O}(\alpha_1 f_1 + \alpha_2 f_2) = \alpha_1 \mathcal{O}(f_1) + \alpha_2 \mathcal{O}(f_2),$$

where f_1 and f_2 have common domain I, and α_1 and α_2 are constant functions.

The differential operator D is linear, for

$$D(\alpha_1 f_1 + \alpha_2 f_2) = D(\alpha_1 f_1) + D(\alpha_2 f_2) = \alpha_1 Df_1 + \alpha_2 Df_2. \qquad (2.2.1)$$

For second-order linear equations, a particular operator L is defined as

$$L = (D^2 + pD + q),$$

where

$$L(f) = D^2 f + p\, Df + qf.$$

The equation in terms of function values is

$$L(f)(x) = D^2 f(x) + p(x)\, Df(x) + q(x)\, f(x) \quad \text{for} \quad x \in I,$$

a common domain of p, q, and f. Following common usage, the last equation is written

$$L(y) = y'' + p(x)y' + q(x)y.$$

In this notation, the general second-order linear equation becomes

$$L(y) = r(x).$$

The equation $L(y) = 0$ is called the second-order *homogeneous** differential equation. The demonstration that L is *linear* is reserved for Exercise 3 [2.2.1].

---

*Note that the meaning of "homogeneous" here is similar to that of *homogeneous linear algebraic equation*; that is, each term contains y or a derivative to the first degree. The word "homogeneous" as used in Section 1.18 is quite different in meaning.

EXERCISES [2.2.1]

. . . .

1. Show that operators $\mathcal{O}_1$ and $\mathcal{O}_2$ are not commutative under composition. Are they commutative relative to addition?

2. Which properties of differentiation were used in establishing the linearity of D^1? See equation (2.2.1).

3. Show that L is linear.

4. Let f_1, f_2, f_3, f_4 be functions on the common domain of $p, q,$ and r. If $L(f_1) = L(f_2) = 0$ and $L(f_3) = L(f_4) = r$, what can be said about $L(kf_1)$, $L(f_1 - f_2)$, $L(f_3 - f_4)$, $L(f_1 + f_3)$?

5. Use the results of Exercise 4 to propose theorems concerning solutions of the equations $L(y) = 0$ and $L(y) = r(x)$.

. . . .

Among the theorems that might have been proposed in Exercise 5 [2.2.1], three are particularly useful and are numbered for future reference.

Theorem 2.2.1

If f_1 and f_2, with common domain I, are solutions of $L(y) = 0$, then $c_1 f_1 + c_2 f_2$ is a solution of $L(y) = 0$, where the c_i are constant functions.

Thus any constant multiple of a solution of $L(y) = 0$ is again a solution. And any sum of solutions is again a solution. Both properties are included in the statement: any linear combination of solutions is a solution. A *linear combination of n functions on I* is the finite sum

$$c_1 f_1 + c_2 f_2 + \cdots + c_n f_n = \sum_{k=1}^{n} c_k f_k,$$

where the constant functions c_k and the f_k are defined on I. This property has far-reaching implications and will be used frequently hereafter. For certain choices of solutions f_1 and f_2, the set $\{c_1 f_1 + c_2 f_2\}$ includes all solutions of $L(y) = 0$; then

$$y(x) = c_1 f_1(x) + c_2 f_2(x)$$

is called the *general solution* of the equation. The special property required is that f_1 and f_2 be linearly independent. This algebraic concept is discussed in the next section.

Theorem 2.2.2

If f_1 and f_3, with common domain I, are solutions of $L(y) = 0$ and $L(y) = r(x)$, respectively, then

$$y(x) = f_1(x) + f_2(x)$$

is a solution of $L(y) = r(x)$.

Theorem 2.2.3

If f_3 and f_4, with common domain I, are solutions of $L(y) = r(x)$, then

$$y(x) = f_3(x) - f_4(x)$$

is a solution of $L(y) = 0$.

The last two theorems are useful in obtaining and classifying the solutions of $L(y) = r$. In particular, Theorem 2.2.2 suggests that solutions of the nonhomogeneous equation may be obtained by taking a particular solution and adding it, in turn, to all the solutions of the related homogeneous equation. Theorem 2.2.3 implies that all solutions of the nonhomogeneous equation may be expressed in that form. Therefore a reasonable procedure would be to first obtain linearly independent solutions to the homogeneous equation; then one particular solution of the nonhomogeneous equation yields all other solutions. But first the concept of linearly independent functions must be considered.

EXERCISES [2.2.2]

. . . .

1. a) Give formal proofs of Theorems 2.2.1, 2.2.2, and 2.2.3.

    b) Do these theorems depend on the order or linearity of the differential equation, or both?

2. Write in operator form the third-order linear equation,

    $$y''' + p_1(x) y'' + p_2(x) y' + p_3(x) y = 0.$$

    Is the operator linear? Conjecture theorems similar to 2.2.2 and 2.2.3.

. . . .

2.3 LINEARLY INDEPENDENT FUNCTIONS

Two functions f_1, f_2 are said to be *linearly independent on the interval I* if there do not exist constants c_1, c_2 (not both zero) such that

$$c_1 f_1(x) + c_2 f_2(x) = 0 \qquad \text{for all} \quad x \in I.$$

Conversely, two functions f_1, f_2 are said to be *linearly dependent on I* if there exist two constants c_1, c_2 (not both zero) such that

$$c_1 f_1(x) + c_2 f_2(x) = 0 \qquad \text{for all} \quad x \in I.$$

For example, the identity function $i: x \to x$ and $2i: x \to 2x$ are dependent on the unit interval $[0, 1]$ because the constants $c_1 = 2$ and $c_2 = -1$ satisfy the equation $c_1 x + c_2 2x = 0$ for $x \in [0, 1]$. The constant function 1_f and the identity function are independent on $[0, 2]$ because there is no nontrivial solution of the equation

$c_1 + c_2 x = 0$. For, at $x = 0$, c_1 must be zero. But if $c_1 = 0$ and $c_2 \neq 0$, then the equation $c_1 1 + c_2 x = 0$ is not satisfied at $x = 2$.

If f_1 and f_2 are linearly dependent and $c_1 \neq 0$, then

$$f_1(x) = -\left(\frac{c_2}{c_1}\right) f_2(x).$$

That is, f_1 is a multiple of f_2. In the case of three or more functions, linear dependence implies that one is a linear combination of the others.

EXERCISES [2.3.1]

. . . .

1. Show that the zero function and any function f on I are linearly dependent on I.
2. Show that
$$f_1(x) = 2 \sin^2 x \quad \text{and} \quad f_2(x) = 1 - \cos^2 x$$
   are dependent on R.
3. Find two different pairs of linearly independent functions.
4. Show that
$$f_1 : x \rightarrow x^2$$
   and
$$f_2 : \begin{cases} x \rightarrow 0 & \text{for} \quad x \geq 0, \\ x \rightarrow x^2 & \text{for} \quad x \leq 0 \end{cases}$$
   are dependent on the interval $(0, \infty)$ or $(-\infty, 0)$ but linearly independent on R.
5. Are the sine and cosine functions independent on $(-\pi/2, \pi/2)$? [*Hint:* Try to show that the assumption of dependence, or independence, leads to a contradiction.]
6. State a definition for the linear dependence of three functions, and then n functions. Show that each definition implies that one of the functions is a linear combination of the others. Can the implication be reversed?

. . . .

2.4 THE WRONSKIAN DETERMINANT

Two functions can be shown to be linearly dependent by finding two constants that "work." But suppose the constants are not easy to find. Should the search be continued? Perhaps the constants do not exist; that is, the functions may be linearly independent. Sometimes the assumption of dependence leads to a contradiction (see Exercise 5 [2.3.1]). But these arguments are often cumbersome, particularly for n ($n > 2$) functions. A determinant, named after the Polish mathematician Wronski (1778–1853), provides a practical test for the linear independence of differentiable functions.

The *Wronskian determinant of two functions* on an interval I is

$$W(f_1, f_2) = \begin{vmatrix} f_1(x) & f_2(x) \\ f_1'(x) & f_2'(x) \end{vmatrix}, \quad x \in I.$$

For the immediate purposes, it is sufficient to know that the second-order determinant

$$\begin{vmatrix} a & b \\ c & d \end{vmatrix} = ad - bc.$$

The Wronskian determinant could be written $f_1(x)f_2'(x) - f_2(x)f_1'(x)$. But determinants generalize easily and are convenient, even for second-order equations.*

Linear independence is an algebraic concept; the theorems and proofs that relate the Wronskian and linear independence deal with algebraic concepts. It becomes increasingly artificial to keep the various branches of mathematics separate. There is, in fact, considerable interplay. Here, the algebraic condition of linear independence leads to general solutions of linear differential equations. Moreover, certain types of linear differential equations may be solved by algebraic methods.

Determinants are tools for solving systems of linear algebraic equations. The transition from the equations to the associated determinant, or vice versa, is the essential idea in the proofs of Theorems 2.4.1 and 2.4.2 relating the Wronskian and linear dependence. Exercises [2.4.1] present the necessary relations.

EXERCISES [2.4.1]

· ·

1. a) Show that if the number pair (x, y) satisfies

$$\begin{cases} a_1x + b_1y = c_1 \\ a_2x + b_2y = c_2 \end{cases} \quad \text{where } a_i, b_i, c_i \in R,$$

then

$$\begin{vmatrix} a_1 & b_1 \\ a_2 & b_2 \end{vmatrix} x = \begin{vmatrix} c_1 & b_1 \\ c_2 & b_2 \end{vmatrix} \quad \text{and} \quad \begin{vmatrix} a_1 & b_1 \\ a_2 & b_2 \end{vmatrix} y = \begin{vmatrix} a_1 & c_1 \\ a_2 & c_2 \end{vmatrix}.$$

b) Under what conditions is (x, y) well defined?

c) If

$$\begin{vmatrix} a_1 & b_1 \\ a_2 & b_2 \end{vmatrix} \neq 0$$

is there more than one solution?

---

*A concise treatment of matrices and determinants is found in Munkres, J. R., *Elementary Linear Algebra*, Reading, Mass.: Addison-Wesley, 1964 (paperback).

d) Discuss the set of solutions $\{(x, y)\}$ if

$$\begin{vmatrix} a_1 & b_1 \\ a_2 & b_2 \end{vmatrix} \neq 0$$

and

i) c_1, c_2 are not both zero,
ii) c_1, c_2 are both zero.

e) Discuss the set of solutions $\{(x, y)\}$ if

$$\begin{vmatrix} a_1 & b_1 \\ a_2 & b_2 \end{vmatrix} = 0$$

and

i) c_1, c_2 are not both zero,
ii) c_1, c_2 are both zero.

2. Each linear equation $a_i x + b_i y = c_i$, $i = 1, 2$, describes a line.

a) What does the zero or nonzero value of

$$\begin{vmatrix} a_1 & b_1 \\ a_2 & b_2 \end{vmatrix}$$

imply about the two lines?

b) Use lines to illustrate geometrically your discussion in 1 (d) and (e).

. . . .

If

$$\begin{vmatrix} a_1 & b_1 \\ a_2 & b_2 \end{vmatrix} \neq 0,$$

then the system

$$\begin{cases} a_1 x + b_1 y = c_1 \\ a_2 x + b_2 y = c_2 \end{cases}$$

has a unique solution (x, y).

If

$$\begin{vmatrix} a_1 & b_1 \\ a_2 & b_2 \end{vmatrix} = 0,$$

then there exist numbers x, y, not both zero, satisfying

$$\begin{cases} a_1 x + b_1 y = 0 \\ a_2 x + b_2 y = 0. \end{cases}$$

The test for linear independence depends on the following theorem.

Theorem 2.4.1

If f_1 and f_2 are differentiable and linearly dependent on I, then $W(f_1, f_2) = 0$ on I.

Proof: By hypothesis, there exist constants c_1, c_2 such that

$$c_1 f_1(x) + c_2 f_2(x) = 0 \qquad \text{for} \quad x \in I.$$

It follows that

$$c_1 f_1'(x) + c_2 f_2'(x) = 0 \qquad \text{for} \quad x \in I.$$

If the f_i and f_i' are evaluated at some x_0 in I, the two equations become

$$c_1 f_1(x_0) + c_2 f_2(x_0) = 0,$$

$$c_1 f_1'(x_0) + c_2 f_2'(x_0) = 0.$$

Remember that the function values at x_0 are numbers. The constants c_1, c_2 are known to exist but their values are not determined. The simultaneous equations in c_1, c_2 imply that (see Exercise 1a [2.4.1])

$$\begin{vmatrix} f_1(x_0) & f_2(x_0) \\ f_1'(x_0) & f_2'(x_0) \end{vmatrix} c_1 = 0 \qquad \text{and} \qquad \begin{vmatrix} f_1(x_0) & f_2(x_0) \\ f_1'(x_0) & f_2'(x_0) \end{vmatrix} c_2 = 0.$$

Hence if c_1, c_2 are not both zero, $W(f_1, f_2) = 0$ at x_0. But the choice of x_0 is arbitrary; the implication holds for all x in I. Therefore the linear dependence of f_1 and f_2 implies that $W(f_1, f_2) = 0$ on I. ∎

It is, in fact, the contrapositive of Theorem 2.4.1 that gives the desired test.

Corollary 2.4.1

If $W(f_1, f_2) \not\equiv 0$ on I, then f_1, f_2 are linearly independent on I.

Note, first, that this result does not supply a test for linear dependence. It has not been shown that a zero Wronskian implies dependence. Second, concerning the test for linear independence, note a simplification. In order to show that the Wronskian is not identically zero on an interval, it is only necessary to show that it is nonzero at a single point of the interval.

EXERCISES [2.4.2]

. . . .

1. Evaluate $W(f_1, f_2)$ on R for
   a) $f_1(x) = x, f_2(x) = 2x$,
   b) $f_1(x) = x, f_2(x) = 1$.

   Compare with results in Section 2.3.

2. Write a formal proof of Corollary 2.4.1.

3. Apply the test (Corollary 2.4.1) to the pairs of functions in Exercises 1 through 5 [2.3.1]. State the implication in each case.

4. Determine the independence or dependence, on some interval I, of

a) $f_1(x) = x^n$,
   $f_2(x) = x^m$

b) $f_1(x) = x$,
   $f_2(x) = e^x$

c) $f_1(x) = x$,
   $f_2(x) = \ln x$

5. Define a Wronskian for three functions. Determine a set of three linearly dependent functions, and three independent functions on some interval I.

. . . .

Exercise 4 [2.3.1] shows that two functions may be linearly independent on I and yet their Wronskian determinant may be zero on I. Hence a nonzero Wronskian gives a definite answer concerning independence; a zero determinant leaves the question unanswered. Fortunately, there is a complete disjunction if the functions are solutions of a linear differential equation.

Theorem 2.4.2

If $W(f_1, f_2) = 0$ and f_1, f_2 are solutions of $L(y) = 0$ on I, then f_1, f_2 are linearly dependent.

Proof: By hypothesis, for any $x_0 \in I$,

$$\begin{vmatrix} f_1(x_0) & f_2(x_0) \\ f_1'(x_0) & f_2'(x_0) \end{vmatrix} = 0.$$

Hence there exist (see (ii) of Exercise 1e [2.4.1]) constants $\hat{c}_1, \hat{c}_2$, not both zero, satisfying the linear algebraic equations

$$c_1 f_1(x_0) + c_2 f_2(x_0) = 0,$$

$$c_1 f_1'(x_0) + c_2 f_2'(x_0) = 0.$$

By Theorem 2.2.1, the function f_3, defined as

$$f_3(x) = \hat{c}_1 f_1(x) + \hat{c}_2 f_2(x),$$

is also a solution of $L(y) = 0$. Furthermore, since $\hat{c}_1, \hat{c}_2$ satisfy the simultaneous equations above, $f_3(x_0) = 0 = f_3'(x_0)$. And according to Corollary 2.1.1, f_3 is unique and is the trivial solution, that is,

$$f_3(x) = \hat{c}_1 f_1(x) + \hat{c}_2 f_2(x) = 0 \qquad \text{on} \quad I.$$

But this is the definition of the linear dependence of two functions on I. ∎

EXERCISES [2.4.3]

. . . .

1. Can the functions in Exercise 4 [2.3.1] be solutions of an equation of the type $L(y) = 0$?

2. a) State an "if and only if" theorem concerning linear dependence and the Wronskian determinant.

   b) State a similar theorem for linear independence.

. . . .

2.5 A GENERAL FORM FOR SOLUTIONS OF THE HOMOGENEOUS LINEAR DIFFERENTIAL EQUATION

Linearly independent solution functions provide a simple form in which to express all the solutions of the equation $L(y) = 0$.

Theorem 2.5.1.

Let f_1 and f_2 be linearly independent solutions of $L(y) = 0$ on the interval I. Let f_3 be any other solution of $L(y) = 0$ on I.

Then
$$f_3(x) = c_1 f_1(x) + c_2 f_2(x) \qquad \text{for some} \quad c_i \in R.$$

Proof: By Theorem 2.4.2, the hypothesis implies that $W(f_1 f_2) \neq 0$ on I. Thus for some $x_0 \in I$,
$$W(f_1(x_0), f_2(x_0)) \neq 0.$$

The function values $f_3(x_0)$ and $f'_3(x_0)$ are real numbers.
  Hence the simultaneous equations
$$f_3(x_0) = c_1 f_1(x_0) + c_2 f_2(x_0),$$
$$f'_3(x_0) = c_1 f'_1(x_0) + c_2 f'_2(x_0)$$

determine a unique solution $(\hat{c}_1, \hat{c}_2)$. (See Exercise 1 [2.4.1].)

  The function $w(x) = \hat{c}_1 f_1(x) + \hat{c}_2 f_2(x)$ is a solution of $L(y) = 0$ on I. Furthermore, $w(x_0) = f_3(x_0)$ and $w'(x_0) = f'_3(x_0)$. Therefore by Theorem 2.1.1,
$$w(x) = f_3(x) = \hat{c}_1 f_1(x) + \hat{c}_2 f_2(x). \quad \blacksquare$$

Hence, if two linearly independent solutions f_1 and f_2 of $L(y) = 0$ on I can be found, then all possible solutions on I are determined as linear combinations of f_1 and f_2. The general solution is
$$y(x) = c_1 f_1(x) + c_2 f_2(x).$$

EXERCISES [2.5.1]

. . . .

1. Write general solutions for equations (2.1.1) and (2.1.2). Do they include the solutions in Table 2.1?

2. Obtain solutions for the equations in Exercises [2.1.1], as special cases of general solutions.

3. Determine* and sketch solutions of $y'' + y = 0$ having

   a) slope 3 at (0, 0),   b) slope $-\frac{1}{2}$ at (0, 0),

   c) slope 0 at (0, 0),   d) slope 1 at (0, 1),

   e) slope $-\frac{1}{2}$ at (0, 1).

[*Hint:* You may have to sketch sums of functions as in Exercise 5 [1.1.1].]

---

* *Note concerning the sine function.* The constants in the general solution are usually determined by substituting the initial conditions in the general solution and its derivative. In Exercise 3 this involves the derivative of the sine function. The rule $D_x \sin x = \cos x$ holds only for the sine function whose set of number pairs includes (0, 0), ($\pi/2$, 1) and (π, 0). It does not hold for the function with number pairs (30, $\frac{1}{2}$), (90, 1), and (180, 0). To convince yourself, sketch two graphs of sine values, one corresponding to degree measure, the other to radian measure† of angles. In order to visualize slopes, in each case use the same scale on the x- and y-axes, that is, equate one degree unit and one vertical unit, and equate one radian unit and one vertical unit. Sketch derivative curves by estimating slopes on the sine curves. Is $f'(0) = 1$ in both cases? A careful look at the proof of the rule $D_x \sin x = \cos x$ shows the essential idea to be

$$\lim_{x \to 0} \frac{\sin x}{x} = 1;$$

this limit holds only for the sine function associated with radian measure. Logarithmic curves are distinguished by subscripts indicating bases. Since only one sine function is used in mathematics at the level of calculus or beyond, no distinguishing mark is used. The sine function, here, always includes the pair ($\pi/2$, 1).

Originally the sine function was defined for angular measures. But this function, as any other, is completely determined by its set of number pairs; the correspondence need not refer to angle measures. For example, the correspondence might represent a relation between time and distance. Suppose the position of a car on a straight track is given by $s = \sin t$, where the units are seconds and meters. Describe the motion of the car. Include remarks about maximum and minimum distances, velocities, and accelerations.

---

4. *A numerical method for a second-order equation.* Compute and sketch numerical solutions of $y'' + y = 0$ with initial conditions (d) and (e) in Exercise 3 above. Use increments of 0.1 on [0, 6.4]. Obtain solutions with three-decimal accuracy. (If a

---

†A definition of radian measure is given in Appendix B.

computer is not available, use increments of 0.2.) Devise a method of your own, or use a modified Euler method for a second-order equation,

$$y(x + h) = y(x) + hy'\left(x + \frac{h}{2}\right),$$

$$y'\left(x + \frac{h}{2}\right) = y'\left(x - \frac{h}{2}\right) + hy''(x).$$

To start, let

$$y'\left(\frac{h}{2}\right) = y'(0) + \frac{h}{2} y''(0) = y'(0) - \frac{h}{2} y(0).$$

5. Do the solution curves in Exercises 3 and 4 above resemble the curve of a particular function? Can each solution be written in the form $y(x) = A \sin (x + \beta)$? Rephrased, given any c_1 and c_2, is it possible to choose constants A and β so that

$$c_1 \sin x + c_2 \cos x = A \sin(x + \beta) \text{ on } R?$$

a) Determine the maximum value of $f_1(x) = c_1 \sin x + c_2 \cos x$. Determine the maximum value of $f_2(x) = A \sin (x + \beta)$. What relation does this suggest between A and the c_i?

b) $f_1(x)$ may be written

$$f_1(x) = (c_1^2 + c_2^2)^{1/2} \left[\frac{c_1}{(c_1^2 + c_2^2)^{1/2}} \sin x + \frac{c_2}{(c_1^2 + c_2^2)^{1/2}} \cos x \right].$$

Rewrite $f_2(x) = A \sin (x + \beta)$ in terms of $\sin x$ and $\cos x$. What relation does this suggest between β and the c_i? Give a geometric interpretation of the relation, that is, let β be the measure of an angle and let the c_i represent lengths of line segments.

c) Are there constants A, β such that

$$c_1 \sin x + c_2 \cos x = A \cos (x + \beta)?$$

d) What are the advantages of each solution form:
   i) $f(x) = c_1 \sin x + c_2 \cos x$,
   ii) $f(x) = A \sin (x + \beta)$?

e) Outline a proof of the relation

$$c_1 \sin kx + c_2 \cos kx = A \sin (kx + \beta).$$

. . . .

Every solution of the equation $y'' + y = 0$ may be written as

$$y(x) = A \sin (x + \beta).$$

Every function that can be written in the form $f(x) = A \sin (\omega x + \beta)$ is called *sinusoidal*. For the sine function, $A = 1$, $\omega = 1$, and $\beta = 0$. Sinusoidal curves resemble the sine curve. The constants A, β, and ω, identified as amplitude, phase displacement, and angular frequency, compare the essential features of each sinusoidal curve to the sine curve. For convenience in notation, the angular frequency

ω, which indicates the number of complete oscillations on the interval $[0, 2\pi]$, is used rather than the frequency $\omega/2\pi$, which indicates the number of oscillations on $[0, 1]$. Sinusoidal functions describe the simplest type of oscillations.

2.6 THE PENDULUM

Clepsydras required much attention and were subject to large errors just in the time involved in starting and stopping the procedure. The instruments were cumbersome. Running water was not everywhere available.

The next significant timepiece was the pendulum, the first mechanical device to display periodic behavior. Galileo wrote in the "First Day" of his *Dialogues*:*

> ... Thousands of times I have observed vibrations, especially in churches where lamps, suspended by long cords, had been inadvertently set in motion.

And in the "Third Day":†

> ... Imagine this page to represent a vertical wall, with a nail driven into it; and from the nail let there be suspended a lead bullet of one or two ounces by means of a fine vertical thread, *AB*, say from four to six feet long (Fig. 2.1). On this wall draw a horizontal line *DC*, at right angles to the vertical thread *AB*, which hangs about two fingerbreadths in front of the wall. Now bring the thread *AB* with the attached ball into the position *AC* and set it free. First it will be observed to descend along the arc *CBD*, to pass the point *B*, and to travel along the arc *CBD*, till it almost reaches the horizontal *CD*, a slight shortage being caused by the resistance of the air and the string; from this we may rightly infer that the ball in its descent through the arc *CB* acquired a momentum (*impeto*) on reaching *B*, which was just sufficient to carry it through a similar arc *BD* to the same height. Having repeated the experiment many times ...

The *equilibrium position* occurs when the bob hangs directly below the suspension point. Unless an outside force or velocity is applied, the bob will not move from this position. The *amplitude* of a complete swing (back and forth) is the maximum displacement of the bob from the equilibrium position. The *period* of a swing is the time it takes for the bob to swing away from and return to the same position of maximum displacement. Using his pulse as a timer, Galileo noticed that for small amplitudes, the period remained constant, even as the amplitude decreased. This periodicity led Galileo to suggest a pendulum-regulated clock.

Consider a mathematical model of the simple pendulum described in Galileo's *Dialogues*. For simplicity, assume that the thread of length *l* is rigid and weightless.

---

*Galilei, Galileo, *Dialogues Concerning Two New Sciences*, "First Day," page 97, translated by H. Crew and A. DeSalvio. New York: Macmillan, 1914.
†Ibid, "Third Day," p.170.

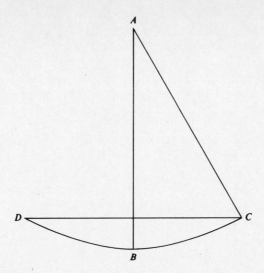

Figure 2.1

Then the suspended object, or bob, must move on a circular arc. Let A be the point of suspension, E the equilibrium position of the bob, and P the position of the bob at any time t, as in Fig. 2.2. Let $\alpha(t)$ be the radian measure of angle EAP; α will be assumed to be a twice differentiable function. Let $s(t)$ be the measure of the arc length $\overparen{EP}$.

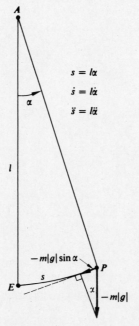

$$s = l\alpha$$
$$\dot{s} = l\dot{\alpha}$$
$$\ddot{s} = l\ddot{\alpha}$$

Figure 2.2

The gravitational force on a bob of mass m is $F_g = -m|g|$. The component of force in the direction of motion is the tangential component, $-m|g| \sin \alpha$. The tangential component of the bob's acceleration is $\ddot{s}(t)$. The model may be restricted to a single function by relating $\ddot{s}(t)$ and the angular acceleration $\ddot{\alpha}(t)$. By the definition of radian measure, $s = l\alpha$ (see Appendix B), and the derivatives enjoy the same proportionality. Newton's law, $F = ma$, may then be written as

$$ml\ddot{\alpha} = -m|g| \sin \alpha \quad \text{or} \quad \ddot{\alpha} + \frac{|g|}{l} \sin \alpha = 0.$$

This is a nonlinear differential equation. It is difficult to solve; the solutions are elliptic integrals.

However, observations have indicated that the period is constant only for small amplitudes. In this region of interest, where $|s|/l = |\alpha|$ is small, $\sin \alpha \doteq \alpha$. A measure of the accuracy is given in Table 2.2. Hence for small $|\alpha|$, the model may be approximated by the linear equation

$$\ddot{\alpha} + \frac{|g|}{l} \alpha = 0.$$

The equivalent equation,

$$\ddot{s} + \frac{|g|}{l} s = 0$$

(obtained by letting $\alpha = s/l$), is often used because it is more convenient to work with distance.

TABLE 2.2

Maximum value of $|\alpha| = |s/l|$ (radians)	Error Maximum value of $|\alpha - \sin \alpha|$
0.07	0.00005
0.14	0.0005
0.30	0.005

This second-order linear equation is similar to some of the examples already considered. Solutions can be obtained by inspection. Note that in the physical situation, the force, and hence the acceleration, is always directed toward the equilibrium position ($s = 0$), that is, the acceleration and displacement always differ in sign. Indeed, this is the statement of the differential equation: the acceleration is proportional to the displacement and opposite in sign.

EXERCISES [2.6.1]

....

1. a) Obtain a general solution for the equation

$$\ddot{s} + \frac{|g|}{l} s = 0.$$

   b) Use the differential equation and simple properties of derivatives to relate the maximum, minimum, and zero velocities and accelerations to specific positions of the bob.

2. a) Let $l = 2.5$ m and $|g| = 9.8$ m/sec². Obtain and sketch solutions satisfying the following initial conditions:

   i) $v_0 = 0$, $s_0 = 15$ cm;
   ii) $v_0 = 15$ cm/sec, $s_0 = 5$ cm;
   iii) $v_0 = -15$ cm/sec, $s_0 = 5$ cm.

   b) What is the amplitude (maximum value of $|s|$) of the motion in each case?
   c) What is the period of the motion in each case?
   d) How may the period be ascertained from the solution function?
   e) Write each solution both as a linear combination and as a sine function. (See Exercise 5 [2.5.1].)

3. Does the initial velocity affect the period or the amplitude of the solution function?

4. a) Write the general solution in terms of v_0 and s_0.
   b) Express the amplitude in terms of v_0 and s_0.
   c) Express β in terms of v_0 and s_0.

5. Construct a simple pendulum and determine the period for different lengths l of the thread.

6. What is the length of a pendulum whose period is 1 sec in Cambridge, Mass: ($|g| = 980.4$ cm/sec²)?

7. How much time would the pendulum of Exercise 6 lose in one day at the equator ($|g| = 978.04$), and in one day on the moon?

8. In 1672, Jean Richter took a pendulum clock from Paris to Cayenne, French Guinea, to make astronomical observations designed to determine the longitude.* He reported with dismay that the clock lost about $2\frac{1}{2}$ min/day in Cayenne. Determine the value of $|g|$ at Cayenne implied by the estimated loss.

9. Starting with the actual value of $|g|$ at Cayenne, estimate the probable loss of Richter's pendulum clock.

....

———————————

*Brown, L. A., "The Longitude," *The World of Mathematics*, Vol. II, pages 780-819. New York: Simon and Schuster, 1956.

2.7 SIMPLE HARMONIC MOTION

If the motion of a pendulum is traced on paper moving in a direction perpendicular to the plane of swing, as in Fig. 2.3, it describes a sinusoidal curve. The solutions of the differential equation are sinusoidal. It is not surprising that periodic motion is described by periodic functions. Periodic motion described by a sinusoidal function is called sinusoidal motion or *simple harmonic motion.**

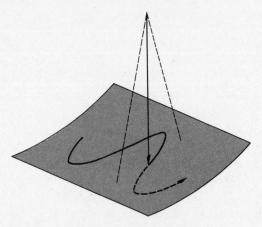

Figure 2.3

Since no resisting force has been considered, the amplitude remains constant for each swing. A and β are determined by the initial conditions $v(t_0) = v_0$ and $s(t_0) = s_0$, and β indicates the displacement of the curve along the time axis. What are the implications of the angular frequency $\omega = \sqrt{g/l}$? Galileo wrote in the "First Day" of his *Dialogues*:†

> ... each pendulum has its own time of vibration so definite and determinate that it is not possible to make it move with any other period [*altro periodo*] than that which nature has given it. For let any one take in his hand the cord to which the weight is attached and try, as much as he pleases, to increase or diminish the frequency [*frequenza*] of its vibrations; it will be time wasted.

2.8 SOME REMARKS ON GRAVITY

When Jean Richter was sent to Cayenne, French Guinea, in 1672, it was thought that g was constant over the earth's surface. Hence a pendulum was expected to keep the same time everywhere. But Richter was so careful in his measurements that he noticed a loss of "about two minutes and a half a day." It was "suspected

---

*A detailed discussion of simple harmonic motion is included in Rogers, Eric M., *Physics for the Inquiring Mind*. Princeton, N.J.: Princeton University Press, 1960.

†Galilei, Galileo, *Dialogues Concerning Two New Sciences*, "First Day," p. 97, translated by H. Crew and A. DeSalvio. New York: Macmillan, 1914.

that this resulted from some error in the observation."* Newton took this so-called error as evidence of the nonspherical shape of the earth. Current measurements show the equatorial radius to be about 22 km longer than the polar radius. Newton had predicted this equatorial bulge as a consequence of the centrifugal force related to the earth's rotation. The resulting acceleration, which is greatest at the equator and zero at the poles, is directed away from the earth and amounts to as much as 3 cm/sec². Actually, g represents the algebraic sum of the gravitational and centrifugal accelerations, or, in practice, the measured acceleration toward the earth's center.† Thus $|g| \doteq 983$ cm/sec² at the pole and 978 cm/sec² at the equator. The latter value is reduced both by the greater distance from the center of the earth and the centrifugal acceleration.

The pendulum, then, can be used (as in Exercise 8 [2.6.1]) to determine the value of $|g|$ at any accessible point. Such measurements have led to the further realization that g also depends on the structure and density of the materials under the earth's surface. Careful maps have been made of the values of g and the directional derivatives, over areas whose substructures are known.‡ In particular, certain geological conditions associated with the accumulation of oil result in specific types of variation in g. Similar patterns of g-values indicate a high probability of oil deposits. Recent contour maps of the gravitational acceleration on the lunar nearside, based on data from lunar orbiter missions, provide new evidence for lunar geologists and historians.

But the very property that makes the pendulum a sensitive instrument in determining g disqualifies it as a portable timepiece. And in the late seventeenth century, a reliable portable clock was wanted for astronomical and navigational measurements. Before analyzing the next development in timepieces—the portable spring clock—consider another instance of simple harmonic motion that might result from the earth's gravitational force.

2.9 TUNNELS THROUGH THE EARTH

Even in the seventeenth century, mathematicians entertained the idea of using gravitational force for transportation in underground tunnels. Today's proposals are serious. First, imagine a tunnel through the earth's center. What is the gravitational force on a body inside the tunnel? For a body on or outside the earth, the

---

*Brown, L. A., "The Longitude," *The World of Mathematics*, Vol. II, page 799. New York: Simon and Schuster, 1956.

†In the *British Astronomical Association Handbook*, 1971, $|g|$ is given as

$$|g| = 978.0310(1 + 0.00530239 \sin^2 \beta - 0.00000587 \sin^2 2\beta - 31.55 \times 10^{-8} h) \text{ cm/sec}^2,$$

where β is the geographic latitude and h the height above sea level measured in meters.

‡Muller, P. M., and W. L. Sjogren, "Mascons: Lunar Mass Concentrations," *Science*, **161**, 3842, 680-684, and cover, August 16, 1968.

force is effectively that of the mass of the earth concentrated at its center. Assume that the earth is spherical. Newton predicted that the force acting on a body inside the earth at a distance r from the earth's center would be that exerted by the mass of a concentric sphere of radius r. The mass outside this sphere does not affect the body. Experiment verifies this prediction.

The following is an intuitive argument. A solid sphere may be thought of as many thin, concentric spherical shells. Consider the effect of one of these shells (Fig. 2.4) at some interior point P. Let P be the nappe of a cone which intersects the

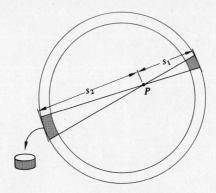

Figure 2.4

shell in two small rings. The curved disks bounded by these rings exert, at P, forces in opposite directions. What is the net force? Let s_1 and s_2 be the distances along the axis of the cone from point P to the shell. The masses of the disks are proportional to the surface areas and hence to the squares of the s_i. On the other hand, the gravitational forces are inversely proportional to squares of the distances s_i. Therefore the two opposing attractions are equal in magnitude and balance each other. The shell can be "covered" with the intersections of cones through a fixed point, and the effect of each pair of disks is zero. Thus the net gravitational force of the shell at an interior point is zero. This argument is vague and includes various approximations, but, hopefully, gives an intuitive feeling for the result. The reader may wish to strengthen the argument or find more rigorous ones.

Our model assumes this result. Let the origin indicate the center of the earth and the tunnel (see Fig. 2.5). Let m_E and r_E be the mass and radius of the earth. Then the gravitational force at a distance r from the center is proportional to the mass of a sphere of radius r, $(r^3/r_E^3)m_E$, and to the reciprocal of r^2. Assuming that only the gravitational force is acting,

$$F = Km_E m r^3 / r_E^3 r^2 = (Km_E m / r_E^3)r.$$

Hence, by Newton's law,

$$|a| = (Km_E / r_E^3)r.$$

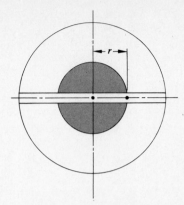

Figure 2.5

Since the direction of the acceleration is toward the origin and differs from r in sign the constant must be negative. The differential equation then is

$$\ddot{r} + (Km_E/r_E^3)r = 0,$$

another model of simple harmonic motion.

A shorter tunnel (see Exercise 3 [2.9.1]) may in fact be practical.

EXERCISES [2.9.1]

. . . .

1. Evaluate the coefficient Km_E/r_E^3 by considering the situation at the surface of the earth, that is, at the opening of the tunnel.

2. Use the implications of the mathematical model to discuss in detail the motion of a body through this tunnel. Include the maximum speed and acceleration attained, the time to go through and the time to return to the starting point.

3. a) Consider the more likely situation of a tunnel along a shorter chord rather than a diameter (see Fig. 2.6). Set up a coordinate system, determine the force at each point

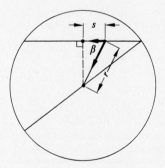

Figure 2.6

in the tunnel, and obtain a differential equation of the motion, assuming that only the gravitational force is acting. [*Hint:* Consider a short tunnel and a diameter tunnel with a common endpoint. Join the centers and use the horizontal component of the gravitational force at *s*.]

  b) Discuss in detail the motion of a body in the short tunnel.

  c) Compare the motion through the two tunnels.

. . . .

SUGGESTED READINGS

Cooper, P. W., "Through the Earth in Forty Minutes," *American Journal of Physics* **34**, 1, 68–70, 1966. The letters concerning this article, in Vol. 34:8, pp. 701–704 make interesting reading.

Edwards, L. K., "High-Speed Tube Transportation," *Scientific American*, **213**, 2, 30–40, August 1968.

Gardner, M., letter, *Scientific American*, **213**, 3, 10, September 1965.

2.10 THE SPRING CLOCK; THE LONGITUDE PROBLEM

The longitude problem demanded an accurate timepiece, one less dependent on local conditions. In 1493, the Pope assigned a certain longitude as the dividing line between the islands belonging to Spain and Portugal. One hundred years later, Spain offered a reward for a practical method of determining longitude. The great exploratory voyages had been made. The European countries had colonies, and ships regularly sailed the open seas. But position at sea was still difficult to determine and shipwrecks were common. The treasures are still being salvaged. When the Spanish plate fleet was wrecked off the Florida Coast, more than one thousand men were lost. Bernard Romans, in *A Concise Natural History of East and West Florida*, published in New York in 1775, described the loss.

> Directly opposite the mouth of the St. Sebastian River happened the shipwreck of the Spanish Admiral who was the northernmost wreck of 14 galleons and a hired Dutch ship, all laden with specie and plate which by (reason) of northeast winds were drove ashore and lost on this coast, between this place and the bleach yard in 1715. A hired Frenchman fortunately escaped, by having steered half a point more east than the others.

These 15 ships, 14 of which still lie five fathoms deep off the Florida coast, were carrying freshly minted gold and silver back to Spain. The total treasure was more than £20 million. The Spaniards salvaged £4 million the following year. Presently a group of Floridians are diving for the treasure and have recovered a value of several million dollars.*

Latitude and longitude together determine position on the earth. Latitude in the northern hemisphere is easily determined by the angle of inclination of the North Star. The idea of longitude is simple. The earth revolves about its axis, that is, through 360°, once in 24 hr. Therefore two longitude lines (great circles through the Poles) differing by 1° represent a solar time difference of $24/360 = 1/15$ hr, or 4 min. A navigator at sea with an accurate timepiece keeping Greenwich solar time, and methods of calculating local solar time (sun fixes and star fixes), can easily determine longitude. But a reliable timepiece was not developed until the latter half of the eighteenth century.

The pendulum was useless at sea. It varied as g varied with position. But even worse, the direction of the gravitational force (that is toward the center of the earth) is not constant with respect to the ship and the pendulum apparatus. As the ship rocks, the pendulum bob moves in a very erratic manner. The basic problem was to develop a portable isochronous timepiece which was not dependent on the gravitational force. The solution was the balance spring, another source of simple harmonic motion.

Mechanical clocks were in existence, even before the pendulum. By means of a series of gears, falling weights were used to turn an hour hand. By 1600, portable clocks driven by the unwinding of a ribbon of steel from a drum had become fashionable. Both types were very irregular. It was remarked by Picard in 1669 that pendulum clocks "marked the seconds with greater accuracy than most clocks mark the half hours." There was no isochronous motion involved in the mechanical clocks, and their accuracy did not approach that needed for determining longitude.

An isochronous regulator was introduced by Robert Hooke (1635–1703) in the form of a balance wheel rotated alternately clockwise and counterclockwise by the harmonic motion of a coiled spring. He also invented an escapement, a gear to transfer the rotational motion of the unwinding steel ribbon into simple harmonic motion. But even with the basic principles in hand, another century of engineering was required to produce a suitable clock. Theoretical principles can have a beautiful simplicity. But physical models must deal with all the small effects that are ignored in mathematical models. First, second, or even third approximations are not enough to deal with motion that must be accurately maintained for many hours and months.

In 1713, the British Parliament offered £20,000 for a clock that would determine longitude within 2 min (34 mi) at the equator. On a six-week voyage, a clock was not to err more than a total of 2 min (3 sec/day). Three major problems prevented

---

*Simpson, Colin, "How the Golden Armada Went Down," *Sunday Times Weekly Review*, p. 41, London, December 11, 1966.
Ibid., Real Eight Inc., *The Sunday Times Magazine*, pp. 8–13, London, December 18, 1966.

the requisite accuracy. The oscillating spring was so regular that changes in frequency were observed during extreme changes in temperature or humidity. The driving force supplied by the unwinding ribbon of steel was conveyed to the balance spring by gears (including the escapement), which introduced considerable friction. Finally, a maintaining force was required to continue the isochronous motion while the clock was being wound. These difficulties were minimized by the clock-maker John Harrison after a lifetime of work. In 1761, his fourth model was tested and showed a loss of 28 mi in longitude after a voyage of more than 5 months. This was well within the specifications, and eventually he received the prize. A detailed description of the Harrison chronometer will be given after a discussion of the coiled spring and equations concerned with general oscillatory motion.

2.11 THE COILED SPRING

Hooke observed that the displacement of a stretched spring is proportional to the force applied. For example, if one end of a coiled spring is fixed (Fig. 2.7), and a

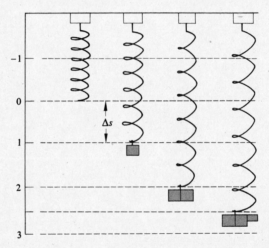

Figure 2.7

load of mass m is attached to the free end, the spring is stretched by the gravitational force, or weight, $|F_g| = m|g|$. Each coil is stretched, but the displacement is the shift Δs of the spring's free end from the unloaded rest position to the loaded rest position. If the mass is doubled, the stretch is doubled:

$$\Delta s \propto |F_g| = m|g|.$$

If H is the constant of proportionality, then $H \Delta s = m|g|$.

When the loaded spring is in equilibrium (at rest), the net force on the load is zero. The gravitational force is balanced by a restoring force F_H of equal magnitude and opposite direction, that is, $F_H = -m|g| = -H \Delta s$. Let s denote a shift in the position of the loaded spring from its equilibrium position $s = 0$ (Fig. 2.8). The total stretch of the spring is $\Delta s + s$; the restoring force is $-H(\Delta s + s)$. The net force on the load, when released, is

$$F = m|g| - H(\Delta s + s) = -Hs.$$

The gravitational force is balanced by the displacement of the equilibrium position; the net force is proportional to the displacement of the load. Rephrased, the restoring force of a loaded spring is proportional to its displacement and opposite in direction:

$$F_H = -Hs.$$

As the concept of force developed, both statements of proportionality, one relating the displacement to the applied force and the other relating the displacement to the

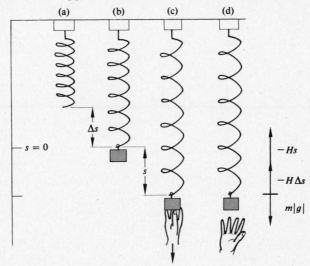

Fig. 2.8 (a) Spring at rest. (b) Loaded spring at rest. (c) Loaded spring stretched a distance s by outside force. (d) Loaded spring released at position s.

restoring force, were referred to as Hooke's law. Of course, Hooke's law holds only on a bounded interval I_H. A spring may be permanently deformed or broken by a large displacement. H and I_H differ for springs of varying materials or cross sections.

    Although the gravitational force is useful in understanding and measuring the restoring force, it does not affect the motion of the loaded spring. Indeed, if the load and spring stiffness are such that the gravitational force is negligible relative to the restoring force, then the spring exhibits the same motion (relative to its axis) whether the axis and displacement are vertical or horizontal. Thus the position of the spring is arbitrary, and the gravitational force is not included in the mathematical models.

In the first model, ignore any resisting force. The condition $ma = F_H = -Hs$, written as a differential equation, is $\ddot{s} + (H/m)s = 0$. Solutions are

$$s = A \sin (\sqrt{H/m}\, t + \beta).$$

The initial conditions must ensure that $A \in I_H$. The motion is simple harmonic. Hooke's law also applies to a torque, or rotational force, on the free end of a spring. The result is a to-and-fro rotation, simple harmonic in terms of the angle of rotation.

This is the third situation that has led to an equation of the form $\ddot{s} + \omega^2 s = 0$, where ω^2 ensures a positive coefficient of s. In each case, ω is determined by the physical situation and is not affected by the initial conditions. That is, the frequency of an oscillating pendulum or spring cannot be altered without changing the physical constants of the system itself, and is called the *natural* or *characteristic frequency*. The equation $\ddot{s} + \omega^2 s = 0$ describes simple harmonic motion of angular frequency ω; any object whose motion is described by this equation is called an *harmonic oscillator*.

The harmonic oscillator is not subject to resisting forces. Doesn't this limit its usefulness as a model? Simple harmonic motion describes oscillatory motion on restricted time intervals, or in systems that include forces which counteract the resistance and maintain a constant amplitude. But the significance of simple harmonic motion lies in its use as the basic model in analyzing more complicated periodic motion. This will be exhibited in the discussions of vibrating strings and membranes (Sections 2.21 and 3.11).

Models including resisting forces would better describe the oscillations of a pendulum or a coiled spring over a long time interval. If the resisting force is proportional to the velocity (as discussed in Section 1.7), the differential equation will still be linear. But the added complications make solution by inspection impractical. How does one find solutions to the equation $L(y) = 0$ if they are not obvious? The method depends on the nature of the functions p and q. The simplest case, as might be expected, occurs when p and q are constant functions.

2.12 HOMOGENEOUS EQUATIONS WITH CONSTANT COEFFICIENTS

Let p and q be constant functions. Let L_c indicate the corresponding operator. The homogeneous equation, then, is

$$L_c(y) = 0 \qquad \text{or} \qquad y''(x) + p\, y'(x) + q\, y(x) = 0.$$

In particular, since constant functions are continuous on R, the solution functions will be defined on R (see Theorem 2.1.1).

Suppose solutions are determined by the hunt-and-find method; that is, various functions are differentiated and substituted to see whether they satisfy the equation. The exponential function $f(x) = e^{mx}$ is particularly simple because each differentiation introduces the same constant factor m. (For comparison try the sine, tangent, logarithmic, and polynomial functions.) Specifically,

$$L_c(e^{mx}) = e^{mx}(m^2 + pm + q).$$

Since the exponential function is a positive function,

$$e^{mx}(m^2 + pm + q) = 0 \quad \text{if and only if} \quad m^2 + pm + q = 0.$$

Hence $f(x) = e^{mx}$ is a solution of $L_c(y) = 0$ if and only if m is a root of $m^2 + pm + q = 0$. This is a powerful and surprising result; the problem of finding a solution of a differential equation has been reduced to finding the solution of an algebraic equation. $m^2 + pm + q = 0$ is called the *auxiliary equation* of the differential equation $L_c(y) = 0$.

EXERCISES [2.12.1]

. . . .

1. Find solutions of the following differential equations. In each case, determine whether the solutions are linearly independent.
   a) $y'' - 3y' + 2y = 0$ b) $y'' + 2y' - 3y = 0$
   c) $y'' - y' = 0$ d) $y'' - 2y' - y = 0$
   e) $y'' - 4y' + 4y = 0$ f) $y'' - y = 0$

2. Show that the solutions $f_1(x) = e^{m_1 x}$, $f_2(x) = e^{m_2 x}$ of $L_c(y) = 0$ are linearly independent if the real numbers m_i are unequal.

3. Write the conclusion and a proof of the following conjecture: If the auxiliary equation of $L_c(y) = 0$ has two real unequal roots m_1, m_2, then the general solution may be written as . . . .

4. Try to find a pair of linearly independent solutions for an equation $L_c(y) = 0$, whose auxiliary equation has equal roots.

. . . .

Since equal roots lead to identical, and hence dependent, solution functions, the auxiliary equation itself does not lead to the general solution. It is necessary to find another solution, by some other means, in order to obtain a linearly independent pair. After a reasonable amount of trial and error, one might arrive at the solution function $f(x) = xe^{mx}$. Pontryagin motivated this solution in the following way. If e^{mx}, $e^{m_* x}$, $m \neq m_*$, are solutions of $L_c(y) = 0$, then by Theorem 2.2.1

$$g(x) = \frac{e^{m_* x} - e^{mx}}{m_* - m}$$

is also a solution. The

$$\lim_{m_* \to m} \frac{e^{m_* x} - e^{mx}}{m_* - m},$$

by definition, is $D_m e^{mx}$. Hence $f(x) = xe^{mx}$ might be expected to be a solution of $L_c(y) = 0$, and indeed it is.*

EXERCISES [2.12.2]

. . . .

1. Show that if the auxiliary equation has equal roots m_1, m_1, then $f_1(x) = e^{m_1 x}$ and $f_2(x) = xe^{m_1 x}$ are solutions and are linearly independent.

2. Write general solutions of the equations:

   a) $y'' - 4y' + 4y = 0$ b) $y'' - 2y = 0$
   c) $y'' + 2y' + y = 0$ d) $y'' + 2y' + 2y = 0$

3. State a conjecture concerning the general solution of the equation $L_c(y) = 0$ whose auxiliary equation has real equal roots.

. . . .

An auxiliary algebraic equation with real coefficients may also have as roots, the complex conjugates $a \pm bi$ (see Exercise 2(d) [2.12.2]). The corresponding solutions

$$f(x) = e^{(a \pm bi)x} = e^{ax} e^{\pm ibx}$$

are complex-valued functions on R. Is there a real solution? If so, how is it determined? The key to this problem is a beautiful result due to Leonhard Euler (1707-1783), namely

$$e^{ibx} = \cos bx + i \sin bx.$$

This may be justified by showing that the series expansions of the exponential, sine, and cosine functions about the origin satisfy the relation for all $x \in R$. The reader unfamiliar with series is advised to assume Euler's result for the present and to establish it after completing the introduction to series in Chapter 3; in particular, see Exercise 9 [3.8.1]. But even this justification assumes that the series $\sum_{n=0}^{\infty} z^n/n!$ defines the exponential function on the complex plane. Here it seems reasonable to accept this assumption, rather than develop carefully the theory of functions of a complex variable. The results of this chapter, with a few interesting exceptions, may be generalized to complex functions on the complex numbers.

Note that for $bx = \pi$, the result relates four of the most interesting numbers in mathematics, $e^{i\pi} = -1$. The historical interest of the numbers -1, i, π, and e is reflected in the descriptive adjectives: negative, imaginary, and irrational and transcendental. The equation $e^{i\varphi} = \cos \varphi + i \sin \varphi$ is on a Swiss postage stamp commemorating the quarter millenium of Euler's birth. (Euler, like his contemporary Mozart, had to find a royal patron. Although Swiss-born, he spent most of

---

*Pontryagin, L. S., *Ordinary Differential Equations*. Reading, Mass.: Addison-Wesley, 1962.

his time in the Russian court. He was one of the mathematicians who considered tunnels through the earth, but in this instance, his particular model has been superseded.)

Euler's result implies that the solution $f(x) = e^{ax}e^{ibx}$ may be expressed as the sum of a real and an imaginary function:

$$f(x) = e^{ax} \cos bx + e^{ax}i \sin bx.$$

In differentiation, i may be treated as a constant. And L_c is linear. Hence

$$L_c(e^{ax} \cos bx + e^{ax}i \sin bx) = L_c(e^{ax} \cos bx) + iL_c(e^{ax} \sin bx) = 0$$

if and only if

$$L_c(e^{ax} \cos bx) = 0 \quad \text{and} \quad L_c(e^{ax} \sin bx) = 0.$$

Since $f(x)$ is a solution, that is $L_c(f(x)) = 0$, it follows that both of the real-valued functions

$$f_1(x) = e^{ax} \cos bx \quad \text{and} \quad f_2(x) = e^{ax} \sin bx$$

are solutions.

EXERCISES [2.12.3]
· ·

1. Apply the operator L_c to the function $f_1(x) = e^{ax} \cos bx$. Show that f_1 satisfies the equation $L_c(y) = 0$, where the roots of the auxiliary equation are $a \pm bi$.

2. Show that the function $f_2(x) = e^{ax} \sin bx$ satisfies the equation $L_c(y) = 0$, where the roots of the auxiliary equation are $a \pm bi$.

3. Show that the functions $f_1(x) = e^{ax} \cos bx$ and $f_2(x) = e^{ax} \sin bx$ are linearly independent on R.

4. Write a general solution of the equation $L_c(y) = 0$, where the roots of the auxiliary equation are $a \pm bi$. Does the same general solution suffice if $a = 0$ or $b = 0$?

5. Write a general solution for each of the following differential equations. Sketch the particular solutions having slope 1 at $(0, 0)$.

   a) $y'' + y = 0$ b) $y'' + y' + y = 0$
   c) $y'' + 2y' + y = 0$ d) $y'' + 3y' + y = 0$

6. Use the auxiliary equation to determine a general solution for the differential equation for the pendulum (Exercise 1 [2.6.1]).

7. What is the effect of a resisting force of $-0.01v$ (due to air resistance) on the motion of a pendulum? (Resisting forces proportional to the velocity v are discussed in Section 1.7).

   a) Set up the differential equation and determine a general solution.
   b) Sketch and discuss the solution for the constants and initial conditions in Exercise 2a (i) [2.6.1].

c) Determine the time interval between two consecutive maxima of the solution function in part (b). Does this model support Galileo's observations concerning the periodicity?

8. a) Discuss the motion of the pendulum in Exercise 7 (b) if the bob is in oil and the drag is $-1v$.

   b) Is the time interval between successive maxima constant?

9. a) Prove that for a function of the form $f(t) = e^{at} \sin \omega t$, the time intervals between successive maxima (or minima) are constant and equal to the period of the function $f(t) = \sin \omega t$.

   b) What can be said about the time intervals between successive maxima of the general solution in 7(a)?

. . . .

It is now possible to state a theorem concerning the solutions of a homogeneous linear equation of order two with constant coefficients.

Theorem 2.12.1

The general solution of the equation $L_c(y) = 0$ may be written in the form

$$f(x) = e^{ax}(c_1 \cos bx + c_2 \sin bx)$$

if the auxiliary equation has roots $a \pm bi$, $b \neq 0$;

$$f(x) = e^{ax}(c_1 + c_2 x)$$

if the auxiliary roots are a, a;

$$f(x) = c_1 e^{ax} + c_2 e^{bx}$$

if the auxiliary roots are $a \neq b$.

The reader may write a formal proof by using appropriate exercises from [2.12.2] and [2.12.3].

2.13 THE DAMPED OSCILLATOR

Exercises 7 and 8 [2.12.3] illustrate damped oscillatory motion. Consider a coiled spring with restoring force $F_H = -Hs$ and resisting force $F_r = -kv$. The net force is $F = -k\dot{s} - Hs$, and the differential equation is

$$\ddot{s} + (k/m)\dot{s} + (H/m)s = 0.$$

The algebraic equation associated with this second-order linear homogeneous equation with constant coefficients has the roots

$$-(k/2m) \pm (1/2m)(k^2 - 4Hm)^{1/2}.$$

If the discriminant is negative, the general solution may be written in the form

$$s(t) = e^{-(k/2m)t} \left[c_1 \cos \left(\frac{H}{m} - \frac{k^2}{4m^2} \right)^{1/2} t + c_2 \sin \left(\frac{H}{m} - \frac{k^2}{4m^2} \right)^{1/2} t \right]. \qquad (2.13.1)$$

This situation obtains when the drag coefficient k is small relative to H. Equation (2.13.1) describes *damped oscillatory motion*: $e^{-(k/2m)t}$ is the *damping factor*. Note the decrease in amplitude with time. Strictly speaking, the motion is not periodic, since the function values are not repeated exactly, but it is still convenient to indicate the duration of each oscillation. The "period," defined as the time between two consecutive local maxima, is constant (see Exercise 9 [2.12.3]; the motion is isochronous. Note also that the frequency differs from the characteristic frequency; the resisting force lengthens the "period."

If the discriminant is zero and the auxiliary equation has equal roots, the general solution is

$$s(t) = e^{-(k/2m)t}(c_1 + c_2 t).$$

This is not an oscillatory function; the solution curve intersects the t-axis in at most one point. The motion is said to be *critically damped*.

If the discriminant is positive, the solution may be written

$$s(t) = \exp \left[-\left(\frac{k}{2m} \right) t \right] \left\{ c_1 \exp \left[+\left(\frac{1}{2m} \right) (k^2 - 4Hm)^{1/2} t \right] \right.$$

$$\left. + c_2 \exp \left[-\left(\frac{1}{2m} \right) (k^2 - 4Hm)^{1/2} t \right] \right\}.$$

The term with positive exponent dominates. As the ratio k/Hm increases it takes longer and longer for $s(t)$ to approach zero. The motion is said to be *overdamped*. Screen and storm doors are often fitted with tubes containing a spring in oil. The oil is chosen so that the drag coefficient results in overdamped motion close to critical damping. It is not practical to obtain critical damping; there is no allowable margin of error.

EXERCISES [2.13.1]

A spring is displaced 0.2 cm by a force of 5 dynes (a force of 1 dyne produces an acceleration of 1 cm/sec^2 on a 1-g mass; that is, 1 dyne = 1 g cm/sec^2). Assume a mass of 0.1 g concentrated at the free end of the spring. The spring is released when the free end is $\frac{1}{2}$ cm on the positive side of the equilibrium position. Determine the equation of motion, find the period, and sketch the solution curve in the situations stated in Exercises 1 through 3 below. ($\sqrt{10} = \pi$ is a convenient approximation here.)

1. Consider only the restoring force of the spring. (The period is that of the balance spring in most watches, so you may check your answer with the ticking of a watch.)

2. Consider, in addition, a resisting force $F_r = -0.02v$ due to air resistance.

3. Let the spring be immersed in a liquid which results in a resisting force $F_r = -2v$.

4. In Exercises 2 and 3, estimate how long it takes for the amplitude to decrease to 1/10 and 1/100 of its original value.

5. If the spring oscillates for only 1 min and the eye cannot detect oscillations less than 0.05 cm in amplitude, approximate the drag coefficient.

6. Determine the resisting force necessary for critical damping.

7. Discuss in detail the effect of the drag on the period. In particular, what happens as the drag approaches the critical damping value?

*8. Choose two damping factors, one slightly greater and the other considerably greater than that needed for critical damping. Consider the resulting motion in each case.

. . . .

2.14 HARRISON'S CHRONOMETER

In Harrison's chronometer, the harmonic oscillator was a *flat* coiled spring, that is, a spring coiled in the shape of a cone and then flattened into one plane. The inner end is fixed; the outer end is attached to the balance wheel which is free to rotate (Fig. 2.9). The distortion in this case is rotational, resulting in an angular displacement. The friction is minimized so that the motion is oscillatory. The frequency of

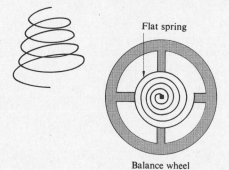

Flat spring

Balance wheel   **Figure 2.9**

the to-and-fro rotation of the balance wheel is determined by the spring constant H, the mass m of the balance wheel, and the drag coefficient k of the system. The coiled spring (called the *balance spring*) is made of two metals which largely compensate each other's changes due to temperature. The driving force is supplied by the mainspring (the unwinding ribbon of metal which can be rewound in its drum every 30 hr). This force is applied to the balance wheel periodically through the escapement and a series of gears. In effect, the balance wheel accepts an impulse, via the escapement, only twice during each period. The small impulse received each time is just enough to negate the damping factor. Thus the coiled spring acting as an isochronous oscillator regulates the rate of unwinding of the mainspring. The ratio of the intermediate gears is such that there is an axle turning at a suitable rate for each of the time hands.

Pendulum clocks were developed along the same lines. The driving force of falling weights was conveyed through gears and an escapement to a regulating pendulum. Most watches today are constructed on the principle outlined above. It is very instructive to take one apart.

Recent advances in timekeeping are based on the theory of resonance (see Section 2.18). Resonance (a particular response of an oscillator to an outside impressed force) is now recognized as a common phenomenon due to the wide applicability of Hooke's law.

2.15 HOOKE'S LAW

A coiled spring is elastic; if distorted, it springs back to its original shape. It is the elastic or restoring force $F_H = -Hs$ that reestablishes the original shape. The endpoints of the interval I_H on which Hooke's law obtains are called the *elastic limits*. Now it is realized that most bodies are elastic to some extent; that is, on some interval of displacements, the body will resume its original shape. For clay, the elastic limit is very small; for certain shapes of metal, the elastic limit is greater than the dimensions of the object. A helical wire spring is one of the most elastic objects; that is, the ratio $F/s = -H$ is essentially constant over a large interval. It is not surprising that the earliest realization of this relation derived from observations of just such a spring. Today, the linear relationship $F = -Hs$ (where F is a distorting or restoring force, s is the displacement, and H is constant) is known as Hooke's law and applies to all objects. It is I_H that determines the degree of elasticity. I_H is usually indicated by a force-displacement curve (see Fig. 2.10) determined by experiment. The linear portion of the curve defines the interval I_H. Outside the interval, Hooke's law does not predict the result of distortions. Displacements beyond the elastic limit will permanently deform or break the object.

Hence any object is an harmonic oscillator; displacements within the elastic limit result in oscillatory motion. Even the earth is an oscillator; its shape is distorted by earthquakes, and the oscillations may last for weeks. Seismographs indicate a fundamental frequency of about 26 oscillations a day. The period is about 54 min; the predicted period for the moon is 15 min. The adjective "fundamental," rather than characteristic, is used because there is a set of characteristic frequencies rather than a single one. (This phenomenon will be discussed relative to the vibrating string and membrane.) The Alaskan earthquake of 1964 caused a permanent deformation in Hawaii; the elastic limit was exceeded. This raises the

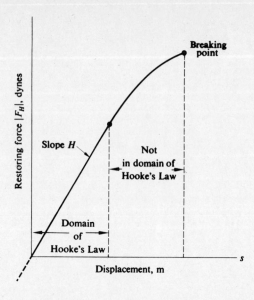

Figure 2.10

question of the role of earthquakes in mountain building and continental drift.*
Earth vibrations are also recorded after hydrogen bomb explosions.

At the other end of the scale consider the force between two atoms (or molecules). The curve showing force as a function of separation distance r (Fig. 2.11)

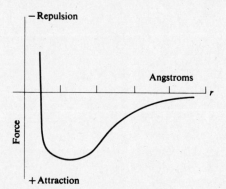

Figure 2.11

is linear in the region where attraction changes to repulsion. Thus an atom in a

---

*Press, Frank, "Resonant Vibrations of the Earth," *Scientific American*, **213**, 5, 28–37, Nov. 1965.
This article may be easier to read after the discussion of characteristic values for the vibrating string
and membrane.

crystal vibrates relative to its neighbors, and is an harmonic oscillator in three dimensions.*

2.16 IMPRESSED FORCES

What happens when a swing, which is a kind of pendulum, is pushed? If the pushes coincide regularly with the swing's motion, the amplitude of each swing increases; indeed, the swing may go over the top. If the pushes are out of step with the swing's characteristic frequency, erratic, rather than oscillatory, motion results.

A tuning fork in oscillation will activate another tuning fork of the same pitch, that is, the same frequency. The first fork impresses a force on the second. This impressed force is the result of periodic increases and decreases in the surrounding air pressure. It is an oscillatory force and is represented by a periodic function. It does not depend on the position or the velocity of the object it affects; it is simply a function of time. This impressed force $F(t)$ corresponds to the $r(x)-$ term in the nonhomogeneous linear equation. The examples above indicate that special effects (called *resonance*) may occur when $F(t)$ has the same period as that of a nearby harmonic oscillator.

Specifically, consider the spring previously discussed. Suppose there is an impressed force $F(t) = A \cos \omega t$, where A and ω are constants. For simplicity, ignore the drag. The resulting differential equation is

$$\ddot{s} + \frac{H}{m} s = \frac{F(t)}{m}.$$

The equations of motion are solutions of a nonhomogeneous differential equation.

2.17 THE NONHOMOGENEOUS EQUATION $L_c(y) = r(x)$ WITH p, q CONSTANT

Theorem 2.17.1

Let r be continuous on the interval I.
Let s_p be a particular solution of the equation $L_c(y) = r(x)$ on I.
Let w be any other solution of $L_c(y) = r(x)$ on I.
Let $f(x) = c_1 y_1(x) + c_2 y_2(x)$ be the general solution of the corresponding homogeneous equation $L_c(y) = 0$.

Then

$$w(x) = c_1 y_1(x) + c_2 y_2(x) + s_p(x) \quad \text{on} \quad I \quad \text{for some} \quad c_1, c_2 \in R.$$

---

*Simple harmonic motion in two dimensions is illustrated by the periodic Lissajoux figures. These may be seen in Tyndall, John, *Sound*, p. 418. New York: Appleton & Co., 1887, and Josephs, Jess J., *Physics of Musical Sound*, Momentum Book 13, p. 14. Princeton, N.J.: Van Nostrand, 1967.

Proof: By Theorem 2.2.3, $y(x) = w(x) - s_p(x)$ is a solution of $L_c(y) = 0$ on I. Hence

$$w(x) - s_p(x) = c_1 y_1(x) + c_2 y_2(x) \qquad \text{for some} \quad c_1, c_2 \in R,$$

and

$$w(x) = c_1 y_1(x) + c_2 y_2(x) + s_p(x) \quad \text{on} \quad I. \quad \blacksquare$$

Thus the general solution of the corresponding homogeneous equation, together with a particular solution of the nonhomogeneous equation, gives a general solution of the nonhomogeneous equation. Since a method for finding solutions of $L_c(y) = 0$ has already been determined, it is only necessary to find a single solution of $L_c(y) = r(x)$. There are various methods, only one of which will be considered here.

In the method of *undetermined coefficients* one constructs a trial function $y = f(x)$ which includes undetermined coefficients, and imposes the condition $L_c(y) = r(x)$ in the hope that the coefficients will be determined and will specify f as a solution function. The trial function must be constructed so that the terms on both sides of the equation $L_c(y) = r(x)$ are similar; only then can the coefficients be compared. Hence f and its derivatives must be like r. If $r(x)$ contains exponential terms, then so must $y(x)$; if $r(x)$ contains sines and cosines, so must $y(x)$. If

$$L_c(y) = (D^2 + 1)y = y'' + y = 4e^{3x},$$

then

$$y(x) = ce^{3x}$$

is a likely solution. The condition

$$L_c(ce^{3x}) = 9ce^{3x} + ce^{3x} = 4e^{3x}$$

is satisfied only if $c = \frac{2}{5}$. Thus the general solution of $y'' + y = 4e^{3x}$ is

$$y(x) = c_1 \cos x + c_2 \sin x + (\tfrac{2}{5})e^{3x}.$$

What if $r(x)$ includes both exponentials *and* sines and cosines? Usually the trial function is a linear combination of the various functions needed. For the equation

$$(D^2 + D + 1)y = y'' + y' + y = e^x + x,$$

a combination such as

$$y(x) = k_0 + k_1 x + k_2 e^x$$

may be necessary. What is a reasonable trial function if $r(x)$ is arctan x or ln x? In these cases, since each successive derivative introduces new terms not in $r(x)$, the method may not work. The procedure is applicable when r has derivatives similar to itself, that is, when r is a linear combination of polynomial, exponential, or sine and cosine functions. Products or composite functions usually increase in complexity when differentiated and do not often lend themselves to this method.

EXERCISES [2.17.1]

. . . .

1. Find particular and general solutions for the following equations.

   a) $y'' + y = 3e^{4x}$ b) $(D^2 - 4)y = 6x^2$

   c) $(D^2 + 2D + 1)y = 4 \sin 2x$ d) $y'' + y' + y = e^x + x$

   e) $(D^2 + 4D + 5)y = x^2 + 3$ f) $(D^2 + 4D + 5)y = e^{-x} + 15$

2. Indicate five functions r in the equation $y'' + p\,y' + q\,y = r(x)$ that are unlikely to yield to the method of undetermined coefficients.

Consider the particular spring and initial conditions specified in Exercises [2.13.1]. Write the differential equation and obtain and sketch solutions in the following situations.

3. No drag; impressed force $F(t) = 2$.

4. No drag; impressed force $F(t) = \sin t$.

5. No drag; impressed force $F(t) = \sin 10t$.

6. Drag $-0.02v$; impressed force $F(t) = 2 \sin 5 \sqrt{10}\ t$.

7. Drag $-2v$; impressed force $F(t) = 2 \sin 5 \sqrt{10}\ t$.

8. Compare and discuss the effects of the impressed forces in Exercises 6 and 7.

9. a) Consider the same spring with an impressed force $F(t) \doteq 2 \sin 5\sqrt{10}\ t$ and no drag. Is the linear combination $k_1 \sin 5\sqrt{10}\ t + k_2 \cos 5\sqrt{10}\ t$ likely to lead to a particular solution of the equation? Explain.

   b) Show that the general solution in Exercise 9(a) may be written in the form

   $$f(x) = c_1 \cos 5\sqrt{10}\ t + c_2 \sin 5\sqrt{10}\ t - 0.63\ t \cos 5\sqrt{10}\ t.$$

   c) Sketch and discuss the solution.

. . . .

2.18 RESONANCE

Exercises 6 and 9(b) [2.17.1] illustrate the phenomenon of *resonance*. For a more general approach, consider first the response of an harmonic oscillator to a periodic impressed force $F(t) = A \cos \omega t$. Here it is convenient to distinguish the natural frequency ω_0 of the oscillator from the frequency ω of the impressed force. Write the differential equation for the motion of the oscillator and show that the general solution may be written in the form

$$s(t) = A_0 \sin (\omega_0 t + \beta) + \frac{A}{m(\omega_0^2 - \omega^2)} \cos \omega t, \qquad \omega \neq \omega_0.$$

The factor $[m(\omega_0^2 - \omega^2)]^{-1}$ determines the magnification of the oscillations of frequency ω. In particular, if ω is close to ω_0, these oscillations are of large amplitude; the resulting condition is called *resonance*.

If $\omega = \omega_0$, show that the solution is

$$s(t) = A_0 \sin(\omega_0 t + \beta) + \frac{A}{2\omega_0 m} t \sin \omega_0 t. \qquad (2.18.1)$$

As t increases, the amplitude of the second term increases without bound. In practice, of course, the oscillations do not build up indefinitely. Increased amplitudes and a constant period imply increased velocities; the drag becomes more significant, and this model no longer approximates the actual motion. When the amplitudes exceed the elastic limit, Hooke's law no longer applies, and the oscillator may be destroyed.

The response of a damped oscillator of natural frequency ω_0 to an impressed force of frequency ω offers a more realistic model. Write the differential equation and show that the solution is

$$s(t) = e^{-(1/2)(k/m)t} A_0 \sin\left[\left(\omega_0 - \frac{k}{4m^2}\right)^{1/2} t + \beta_1\right]$$

$$+ A[m^2(\omega_0^2 - \omega^2)^2 + k^2\omega^2]^{-1/2} \cos(\omega t + \beta_2). \qquad (2.18.2)$$

The first term represents damped oscillatory motion; it is called the transient response and becomes negligible for large t. (This particular solution assumes that the drag coefficient k is small enough to allow oscillations, that is, $k < 2\omega_0 m$. If k is larger, the first term will describe critical or overdamped motion.) For a fixed ω, the second term, called the *steady-state response*, has constant amplitude and describes the motion for large t. The magnification factor is

$$P(\omega) = [m^2(\omega_0^2 - \omega^2)^2 + k^2\omega^2]^{-1/2}.$$

What happens if the impressed frequency ω varies? If ω approaches ω_0, the amplitude of the steady-state response increases, but in this case there is a maximum value [see Exercise 2(b) below]:

$$\frac{A}{k}\left(\omega_0^2 - \frac{k^2}{4m^2}\right)^{-1/2}.$$

The impressed frequency that produces the maximum amplitude is not ω_0, but is close to ω_0 if k/m is small; it is called the *resonant frequency* of the system. The graph $\{(\omega, P^2(\omega))\}$ in the vicinity of the resonant frequency is called the *resonance curve*; $P^2(\omega)$ rather than $P(\omega)$ is used because it is proportional to the energy in the system. The exercises below continue the discussion of the resonance curve.

EXERCISES [2.18.1]

1. a) Use methods of calculus to sketch the resonance curve $P^2(\omega) = [m(\omega_0^2 - \omega^2)]^{-2}$
   for the response of an harmonic oscillator to the impressed force $F(t) = A \cos \omega t$.
   In particular, consider the spring described in Exercises [2.13.1]. Discuss the motion
   for different values of ω as predicted by the resonance curve.
   b) Write the differential equation for the case $\omega = \omega_0$. Obtain the general solution
   (2.18.1).
   c) If, for the spring in Exercise 1(a), the elastic limit is 3 cm and the breaking point is
   5 cm, on what time interval is the solution a reasonable model?

2. a) Write the differential equation for the response of a damped oscillator to the impressed force $F(t) = A \cos \omega t$. Obtain the general solution (2.18.2).
   b) What is the resonant frequency of the system?
   c) Derive the expression for the maximum amplitude, produced by the resonant frequency in terms of w_0, k, and m.

3. a) Sketch the resonance curve $P^2(\omega) = [m^2(\omega_0^2 - \omega^2)^2 + k^2\omega^2]^{-1}$ showing the
   response of a damped oscillator to the impressed force $F(t) = A \cos \omega t$. Consider
   the same spring as in Exercise 1(a); use resisting forces of $-0.02v$, $-0.2v$, and $-2v$.
   b) Discuss the effects of the drag coefficient on the resonance curve.
   c) For what values of ω will the spring be permanently deformed? Assume the same
   constants as stated in Exercise 1(b).

4. The sharpness of a resonance curve is measured by the half-width (Fig. 2.12) of the
   curve (the width of the curve at one-half the maximum height). Find the half-widths of

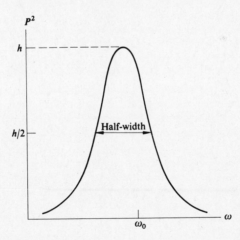

Figure 2.12

the curves in Exercise 3(a). What is the relation between the drag coefficient and the
sharpness of the resonance curve?

The smaller the ratio k/m, the sharper the resonance curve. An impressed force of frequency close to the resonant frequency of a mechanical system, may destroy the system.

2.19 EXAMPLES OF RESONANCE

Galileo described an instance of resonance* that is still used in the ringing of heavy free-swinging tower bells.

> Even as a boy, I observed that one man alone by giving these impulses at the right instant was able to ring a bell so large that when four, or even six, men seized the rope and tried to stop it they were lifted from the ground, all of them together being unable to counterbalance the momentum which a single man, by properly timed pulls, had given it.

Eyewitness accounts of large bridges in resonance are particularly interesting. The rule that troops must break step when crossing a bridge may have originated in England after the collapse of the Broughton suspension bridge near Manchester in 1831. Excerpts from the accounts in the Manchester Guardian and the Manchester Chronicle, both of April 16, 1831, are given below.

The *Manchester Guardian:*

> A very serious and alarming accident occurred on Tuesday last, in the fall of the Broughton suspension bridge, erected a few years ago by John Fitzgerald, Esq., whilst a company of the 60th rifles were passing over it; and, although fortunately no lives were lost, several of the soldiers received serious personal injuries, and damage was done to the structure, which will require a long time and a very considerable expense to repair.
>
> It appears that, on the day when this accident happened, the 60th regiment had had a field-day on Kersall Moor, and, about twelve o'clock, were on their way back to their quarters. The greater part of the regiment is stationed in the temporary barracks, in Dyche-street, St. George's Road, and took the route through Strangeways; but one company, commanded, as it happened singularly enough, by Lieut. P. S. Fitzgerald, the son of the proprietor of the bridge, being stationed at the Salford barracks, took the road over the suspension bridge, intending to go through Pendleton to the barracks. Shortly after they got upon the bridge, the men, who were marching four abreast, found that the structure vibrated in unison with the measured step with which they marched; and, as this vibration was by no means unpleasant, they were inclined to humour it by the manner in which they stepped. As they proceeded, and as a greater number of them got upon the bridge, the vibration went on increasing until the head of the column had nearly reached the Pendleton side of the river. They were

---

*Galilei, Galileo, *Dialogues Concerning Two New Sciences*, "First Day," page 98, translated by H. Crew and A. DeSalvio. New York: Macmillan, 1914.

DREADFUL ACCIDENT AT BROUGHTON SUSPENSION BRIDGE.

Figure 2.13(a)

then alarmed by a loud sound something resembling an irregular discharge of fire-arms; and immediately one of the iron pillars supporting the suspension chains, viz. that which was to the right of the soldiers, and on the Broughton side of the river, fell towards the bridge, carrying with it a large stone from the pier, to which it had been bolted. Of course that corner of the bridge, having lost the support of the pillar, immediately fell to the bottom of the river, a descent of about 16 or 18 feet; and from the great inclination thereby given to the roadway, nearly the whole of the soldiers who were upon it were precipitated into the river, where a scene of great confusion was exhibited. Such of them as were unhurt got out as well as they could, some by scrambling up the inclined plane which the bridge presented, and others by wading out on the Broughton side; but a number were too much hurt to extricate themselves without assistance, which was immediately rendered by their comrades.

... There is no doubt that the immediate cause was the powerful vibration communicated to the bridge by the measured and uniform step of the soldiers. If the same, or a much larger number of persons had passed over in a crowd, and without observing any regular step, in all probability the accident would not have happened, because the tread of one person would have counteracted the vibration arising from that of another. But the soldiers all stepping at the same time, and at regular intervals, communicated, as we mentioned in describing the accident, a powerful vibration to the bridge, which went on increasing with every successive step; and which, causing the weight of the bridge to act with successive jerks on the stay-chains, had a more powerful effect upon them than a dead weight of much larger amount would have had, and at length broke one of the cross-bolts by which the links of the chain are joined together. Perhaps this accident, alarming and injurious as it has been, may have the effect of preventing some more dreadful catastrophe in other quarters.

The *Manchester Chronicle:*

... It has been stated by some scientific men, and we fully concur in the opinion, that the peculiar manner in which the soldiers marched whilst on the bridge had no slight share in causing the accident. Before they reached the bridge we are told that they were walking "at ease," but when they heard the sound of their own footsteps upon it, one or two of them involuntarily began to whistle a martial tune, and they all at once, as if under a command from their officer, commenced a simultaneous military step. This uniform motion naturally gave great agitation to the bridge, the violent effects of which would be most severely felt at each end. As a familiar illustration of our meaning we may remark, that if a rope, the ends of which being fastened to opposite walls, should be much agitated in the centre, its motion would be far more violent at the ends than in any other part.

In 1836, the Brighton chain-pier collapsed for the second time. In this case, the impressed force was the wind. An account and sketch published by Lieut.-Col. Reid in the first volume of the Professional Papers of the Corps of Royal Engineers, is given below.

"The same span of the Brighton chain-pier (the third from the shore), has now twice given way in a storm. The first time it happened in a dark night, and the storm was accompanied by much thunder and lightning: the general opinion of those who do not inquire into the causes of such matters was that it was destroyed by lightning; but the persons employed about the pier, and whose business it was to repair it, were satisfied that the first fracture was neither caused by lightning nor by the waters, but by the wind.

The fracture this year was similar to the former, and the cause evidently the same. This time, it gave way half an hour after mid-day, on the 30th of November 1836, and a great number of persons were therefore enabled to see it.

The upper one of the two sketches annexed, shows the greatest degree of undulation it arrived at before the road-way broke; and the under one shows its state after it broke; but the great chains from which the road is suspended remained entire.

When this span became relieved from a portion of its load by the road-way falling into the sea, its two piers went a little on one side, and the curve of the chain became less, as in the sketch. The second and fourth spans in these sketches, are drawn straight, merely to shew better the degree of undulation of the third span. These also undulated greatly during the storm, but not in the same degree as the third span. A movement of the same kind in the road-way has always been sensibly felt by persons walking on it in high winds; but on the 29th of November 1836, the wind had almost the same violence as in a tropical hurricane, since it unroofed houses and threw down trees. To those who were at Brighton at the time, the effect of such a storm on the chain-pier was a matter of interest and great curiosity. For a considerable time, the undulations of all the spans seemed nearly equal. The gale became a storm about eleven o'clock in the forenoon, and by noon it blew very hard. Up to this period many persons from

Figure 2.13(b)

curiosity went across the first span, and a few were seen at the further end; but soon after midday the lateral oscillations of the third span increased to a degree to make it doubtful whether the work could withstand the storm; and soon afterwards the oscillating motion across the road-way, seemed to the eye to be lost in the undulating one, which in the third span was much greater than in the other three; the undulatory motion which was along the length of the road is that which is shewn in the first sketch; but there was also an oscillating motion of the great chains across the work, though the one seemed to destroy the other, as they did not both (at least as far as could be seen) take place in a marked manner at the same time.

At last the railing on the east side was seen to be breaking away, falling into the sea; and immediately the undulations increased; and when the railing on this side was nearly all gone, the undulations were quite as great as represented in the drawing."

Mr. John Scott Russell, aware of both the Broughton and Brighton catastrophes, published a paper "On the Vibration of Suspension Bridges and Other Structures, and the Means of Preventing Injury from This Cause" in *Transactions of the Royal Scottish Society of Arts*, Vol. 1, 1841. He noted, with great clarity, that a suspension bridge is vulnerable to resonance because, like a taut string, the structure has a set of characteristic frequencies. But his sound suggestions for preventive construction went unheeded in this country. A suspension bridge over

the Ohio River at Wheeling was completed in 1849, and succumbed to resonance
five years later. Technical accounts blamed only the storm and did not recognize
the condition of resonance. But a reporter who witnessed the collapse gave a
vivid description which appeared in the *New York Times* on May 22, 1854.

"For a mechanical solution of the unexpected fall of this stupendous structure, we must
await further developments. We witnessed the terrific scene. The great body of the
flooring and the suspenders, forming something like a basket swung between the towers,
was swayed to and fro like the motion of a pendulum. Each vibration giving it increased
momentum, the cables, which sustained the whole structure, were unable to resist a force
operating on them in so many different directions, and were literally twisted and wrenched
from their fastenings . . . "

In 1940, The Tacoma Narrows Bridge collapsed in Puget Sound, Washington,
six days after the official opening. The motion of the bridge had attracted much
attention locally, even during construction. Professor H. C. Carver of the University of Michigan had spent the summer on the West Coast, and predicted to his
class in probability and statistics in September 1940 that the bridge would collapse
that autumn. On the fateful day there was a 40-mph wind. Many people came to
watch, some with movie cameras, so the buildup of the oscillations was recorded on
film.* It is now established that a steady wind against solid girders of a flexible
structure may produce a periodic force. Several features of the Tacoma Bridge
made it susceptible to resonance: the solid girders, the narrowness of the span and
the coincidence of the natural frequencies of the tower and spans. Interesting
details of this incident and some of the previous ones are discussed in *Bridges and
Their Builders* by David Steinman and Sara Ruth Watson, pp. 356–357, Dover
Publications, New York, 1957.

In 1959 and 1960 several Electra turboprop planes crashed. They seemed to
"explode" in midair. The investigations of the Civil Aeronautics Board indicated
that the disintegration of the planes was due to mechanical resonance. In particular,
one of the machine packages, when not fastened securely, oscillated at a frequency
within the peak of the resonance curve of the wing. These oscillations, about 3/sec,
acted as an impressed force on the wing. Within 30 sec, the amplitude of the wing's
vibrations increased beyond the elastic limit. After the wing had broken and was
torn off, the rest of the plane disintegrated rapidly. Small models with loose
machine packages were tested in a wind tunnel. Precisely this type of disintegration
was observed. The planes were modified by altering the machine package to
change its natural frequency of vibration.†

---

*Film loop 80-218, Tacoma Narrows Bridge Collapse, is available from Ealing Film Loops,
Cambridge, Mass.

†See *New York Times*, May 22, 1960, page 1 of the Travel Section. The whole story of the crashes,
charges, political implications, and investigations can be followed by consulting the *New York
Times Index* for 1959, 1960, 1961, under "Airlines, U.S., accidents."

In the 1969 trial run of the Queen Elizabeth II, a resonance condition caused the blades of the turbine rotors to snap. The ship was refused by Cunard Lines, and several months of sailings had to be canceled.

Resonance can be a useful phenomenon as evidenced, for example, by linear electrical circuits. There is a complete analogy: the resistance R of a wire is equivalent to the drag coefficient; the inductance L corresponds to mass; Hooke's constant becomes the reciprocal of the capacitance C; the voltage is the impressed force, and the charge q and current $\dot{q} = I$ correspond to position and velocity. Equivalent conditions result in the equation

$$L\ddot{q} + R\dot{q} + \frac{1}{C} q = V(t).$$

Hence for small resistances, the circuit is, in fact, an electrical oscillator, and all that is known about mechanical oscillators applies to electrical oscillators. In practice, the analogy is more often used the other way. It is easier to build circuits and change the electrical constants than to build and alter mechanical systems. Therefore tests are often run on electrical oscillators to determine performance and design specifications for mechanical systems. Certain aspects of automobile design are determined in this fashion.

A familiar application of resonance is the radio tuning device. Each radio station broadcasts at its own special frequency. All broadcast waves offer a periodic impressed force to the receiving circuit of a radio. How is a single frequency selected? The tuning knob changes the capacitance or inductance—and hence the natural frequency w_0— of the receiving circuit. When the frequency of a particular station falls within the domain of the resonant frequencies of the circuit, the amplitude of that particular frequency increases, that is, resonance occurs. Hence that particular station may be heard, while all the other signals remain below the audible level. In the same way, the receiver used with a radio telescope can be tuned to detect signals in the vicinity of a particular wavelength. If tuned to the region of 21 cm, slight differences in wavelength (due to the Doppler effect) indicate the velocity of the corresponding hydrogen cloud. (See the discussion following Exercises [1.18.1].)

2.20 RESONANCE IN TIME MEASUREMENT

An electrical oscillator is an efficient and simple timekeeper. Household current is the isochronous regulator for the usual electric clock. The alternating current plays the same role as the flat spring in the watch or the pendulum in the grandfather clock. The system of gears is replaced by a series of calibrated circuits.

Electrical circuits are very flexible. By imposing suitable constants, very high frequencies may be realized. A high-frequency oscillator may be calibrated to a series of circuits of progressively lower frequencies. With the aid of a high-speed counting device, such a system may be used to measure very small time intervals. But there is, of course, some variation in the natural frequency of the electric circuit; the imposed constants are affected by external conditions. Hence there is a lower bound to the time intervals that can be measured with precision in this manner. For intervals of microseconds (10^{-6} sec) or less, isochronous regulators of high frequency and greater stability are needed.

A quartz crystal has a steadier vibration frequency. When placed in a proper circuit, the crystal resonates in response to the electric oscillations and imposes its steady frequency on the circuit. The error is less than one part in 10^6. These crystal clocks, located in observatories throughout the world, agree with one another, but disagree with the earth as a timekeeper. Corrections must be made for such effects as the earth's wobble. But, in addition to the predictable differences, there are unexplained irregularities in the earth's rotations. In order to understand these small irregularities, it is necessary to make finer time measurements. The accuracy of the quartz crystal is limited by the tendency of its frequency to shift with temperature and age.

More recent developments depend on atomic vibrations as the isochronous regulators.* Within the ammonia molecule, the nitrogen atom oscillates with a frequency of 2.387×10^{10} cycles/sec. The ammonia clock is a refined quartz clock. The resonance of the ammonia molecules is used to correct the irregularities in the crystal's oscillations. Electrical oscillations are impressed on the crystal. The particular frequency chosen (within the crystal's domain of resonance) is such that when multiplied electrically (that is, calibrated to electrical oscillators of higher frequency) and converted to radio waves, it offers an impressed force of 2.387×10^{10} cycles to ammonia gas enclosed in a waveguide. The resonance curve of ammonia is sharp. A small variation in the impressed frequency reduces the vibrations of the ammonia molecules, and a feedback mechanism corrects the impressed frequency. The error is reduced to one part in 10^8.

The phenomenon may be described in different terms. Energy is supplied to the ammonia gas. If the nitrogen atoms within the molecules resonate, the molecules are said to absorb the energy. The *absorption curve*, which indicates the power output from the waveguide of ammonia gas as a function of the frequency of the energy input, is another representation of the resonance. The power output is low when the molecules are resonating and high otherwise.

Even higher precision is obtained with cesium; its natural frequency is 0.9192×10^{10}. Cesium atoms exposed to this frequency change their energy states. The resonance curve of cesium is sharper than that of ammonia. The cesium clock uses the resonance of cesium atoms to correct the oscillations of a quartz crystal; the error is less than one part in 10^{10}. Atomic clocks are being further improved by treating the atoms in such a way that the resonance curve is sharpened. Intervals of 10^{-12} sec can now be measured.

---

*Lyons, Harold, "Atomic Clocks," *Scientific American*, **196**, 2, 71–82, Feb. 1957.

Compared to this order of time measurement, the earth's rotation is not periodic. At present, every day is about 2×10^{-3} sec longer than the previous day. Even a century ago, G. H. Darwin predicted the lengthening of the day as a result of the tides. But time measurements then were too crude to offer supporting evidence.

The tidal consequences . . . are emphatically of this non-periodic class—the day is always lengthening, the moon is always retreating. Today is longer than yesterday; tomorrow will be longer than today. It cannot be said that the change is a great one; it is indeed too small to be appreciated even by our most delicate observations.*

Now there is ample evidence. Fossil corals with daily growth bands indicate about 400 days per year in the mid-Devonian period (3.7×10^8 years ago), and about 428 days at the beginning of the Cambrian period (5.7×10^8 years ago). Living corals deposit about 360 bands a year. The secretion rate of calcium carbonate changes from day to night.†

Present measurements of the earth's period also indicate unpredictable irregularities of the same order of magnitude (10^{-3} sec). Since the day is not constant, neither is the second, which is 1/86,400 of a day. In 1956, the second was defined as 1/31,556,925.9747 of the tropical year 1900. (A tropical year is the time from one equinox to the return of that equinox.) In 1967, the second was redefined as 9,192,631,770 cycles of the frequency associated with the transition between two energy levels of the isotope cesium-133. (In 1960, the meter was redefined as 1,650,763.73 wavelengths in a vacuum of the reddish-orange radiation emitted by the transition between the energy levels $2p_{10}$ and $5d_5$ of the krypton-86 atom.‡)

So far, all the models of periodic motion have been simple harmonic or slight variations thereof. Most oscillations are not that simple; it would be misleading to leave such an impression. The wave equation serves as a model for various physical situations where the oscillations are more complicated, and at the same time introduces Fourier series. Here the wave equation is developed as a model of the motion of a plucked string.

2.21 THE PLUCKED STRING

Not all periodic functions are sinusoidal. Not all oscillations represent simple harmonic motion. For example, a single musical note played on an oboe may activate several piano strings or tuning forks of different frequencies. Three periods

---

*Ball, Sir R. S., *Time and Tide*, p. 74. London: Society for Promoting Christian Knowledge, 1889.
†Clancy, E. P. *The Tides*, Science Studies Series, Anchor S56. Garden City, N.Y.: Doubleday, 1968. Runcorn, S. K. "Corals as Paleontological Clocks," *Scientific American*, **215**, 4, 26-33, Oct. 1966.
‡Astin, Allen V., "Standards of Measurement," *Scientific American*, **218**, 6, 50, 50-62, June 1968: Lord Ritchie-Calder, "Conversion to the Metric System," *Scientific American*, **223**, 1, 17-25, July 1970.

of a sound wave generated by an oboe sounding middle C are shown in Fig. 2.14.

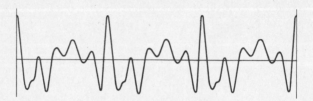

Figure 2.14

Consider the impressed force

$$F(t) = A_1 \cos \omega_1 t + A_2 \cos \omega_2 t$$

acting on an oscillator. Corresponding to each cosine function there is a solution of the equation $\ddot{s} + \omega^2 s = F(t)/m$. The sum of these solutions represents the response of the system to both forces applied simultaneously. The steady-state response is the sum of two cosine functions, and similarly for any number of forces. One of the beautiful results in mathematics is that of Fourier (1768–1830). It states that practically all periodic functions may be represented as the sum of sine and cosine functions. That is, periodic motion is the result of simultaneous simple harmonic motions of different phases and amplitudes, and a common period. There are even practical methods for sorting out the component parts, somewhat like the separation of radio frequencies by resonance.

In order to demonstrate nonsimple vibrations, the motion of a vibrating string will be considered. This example has several other points of interest. It introduces a partial differential equation with boundary conditions; it demonstrates a common method of solving such equations, namely, by reducing them to ordinary differential equations; and since the ordinary equations in this case are linear, it provides an illustration of the use of linear equations in solving partial differential equations.

Consider a taut string of length L, with both ends fastened securely. On the violin, this would be the section of string between the nut and the bridge. Suppose a point of the string is pulled to the side (see Fig. 2.15) and then released, as in plucking. What is the resulting motion? It is readily observed that the string vibrates and that the amplitudes are very small relative to the length of the string.

Path of motion

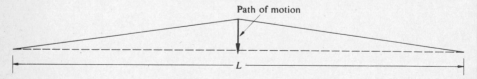

Figure 2.15

(A large displacement will break a taut string.) As usual, a number of simplifying assumptions have to be made in order to obtain a workable mathematical model.

The motion of particular points, rather than that of the whole string, will be considered. It will be assumed that the string vibrates in a single plane, and further, that each point moves along a line perpendicular to the equilibrium position of the string.

The only force considered will be that due to the tension in the string. (For a taut string, the gravitational force is insignificant relative to the tension.) At each point there will be two forces, each of magnitude T, acting in opposite directions along the string and away from the point. T is assumed to be constant for all points throughout the motion.

Let one endpoint be the origin, and the other be on the positive x-axis (see Fig. 2.16). Then a point on the string with equilibrium position $(x, 0)$ has, during the vibration, position (x, y) where y depends on t and x.

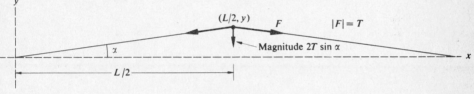

Figure 2.16

As a first approximation, consider one particular point such as the midpoint. The usual assumption, as in the previous applications, is that the mass is concentrated at the point in question. The total force acting in the direction of motion, that is, vertically, is $-2T \sin \alpha$. Note that the net force is negative when α is positive. For small amplitudes, $\sin \alpha$ may be approximated as $y/(L/2)$. Thus Newton's law leads to the differential equation $\ddot{y} + (4T/mL)y = 0$. This equation is an "old friend" and the motion is well determined. The point vibrates back and forth with a period of $\pi(mL/T)^{1/2}$ cycles per second. However, the period of such a string has long been observed and measured, and is known to be approximately $2(mL/T)^{1/2}$. This sinusoidal model gives a period which is off by more than 50%. Since the frequency and period are crucial to any oscillatory motion, it seems futile to pursue this model further. A careful look at the assumptions may suggest a better approximation.

2.22 THE WAVE EQUATION

In all the previous applications, the moving body could be considered as a point relative to the rest of the system; reasonable models were obtained by considering the mass concentrated at this point. In this instance, the mass is distributed uniformly along the entire length of the string which is itself the whole system. It seems more reasonable to consider a small piece of the string whose endpoints have x-coordinates x_i and x_{i+1} (Fig. 2.17). The motion through the point $(x_i, 0)$ may

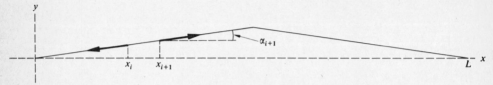

Figure 2.17

then be considered as the limiting case of the motion of the segment as x_{i+1} approaches x_i. The disadvantage of this second approach is that it leads to a more difficult equation involving partial derivatives. The advantage is that the solutions agree with observations and measurements, and provide a basis for the theory of music.

In this second model, the concentrated mass is replaced by the assumption that the string has a uniform linear density σ. The mass of the segment in equilibrium position is, then, $\sigma(x_{i+1} - x_i)$. The assumption that the mass of the segment remains constant throughout the motion is not unreasonable for small amplitudes and slopes. The effective point of motion and concentration of mass is taken as the centroid of the segment. Hopefully, the motion through each point $(x, 0)$ may be described by some function $y = f(x, t)$, whose first and second partial derivatives exist and are continuous. In particular, consider the partials $\partial y/\partial x = y_x$ and $\partial y/\partial t = y_t$. For a fixed t_*, $y_x(x_a, t_*)$ indicates the slope, at x_a, of the string in its position at time t_*. At a fixed x_*, $y_t(x_*, t_a)$ indicates the vertical velocity at time t_a of the point with first coordinate x_*. What is the physical interpretation of

$$\frac{\partial^2 y}{\partial x^2}(x_a, t_*) = y_{xx}(x_a, t_*) \qquad \text{and} \qquad \frac{\partial^2 y}{\partial t^2}(x_a, t_*)?$$

The forces acting on each segment of the string at time t may be represented by two force vectors of constant magnitude T. Each vector is directed away from the segment, along the tangent to the string at $(x_i, y(x_i, t))$. Let $\alpha_i = \alpha(x_i, t)$ indicate the measure of the angle from the horizontal to the force vector at $(x_i, y(x_i, t))$. For small α_i the vertical component of the force, $T \sin \alpha_i$, may be reasonably approximated as

$$T \tan \alpha_i = T \frac{\partial y}{\partial x}(x_i, t).$$

Then the net force, $T \sin \alpha_{i+1} + T \sin \alpha_i$, is

$$T \frac{\partial y}{\partial x}(x_{i+1}, t) - T \frac{\partial y}{\partial x}(x_i, t).$$

(Note that α_i at the left-hand endpoint is close to π. In this region, the sine and tangent functions have approximately the same magnitudes but differ in sign.) For a fixed t, and $x \in [0, L]$, $\partial y/\partial x$ defines a function of one variable. The mean value theorem asserts that for some $\bar{x}_i$ between x_i and x_{i+1},

$$T \frac{\partial y}{\partial x}(x_{i+1}, t) - T \frac{\partial y}{\partial x}(x_i, t) = T \frac{\partial^2 y}{\partial x^2}(\bar{x}_i, t)[x_{i+1} - x_i].$$

What assumptions must be made to ensure that the theorem applies? At the centroid,

$$ma = \sigma[x_{i+1} - x_i] \frac{\partial^2 y}{\partial t^2}\left(\frac{x_i + x_{i+1}}{2}, t\right).$$

Newton's law gives the relation, for any time t and an interval $[x_i, x_{i+1}]$,

$$T \frac{\partial^2 y}{\partial x^2}(\bar{x}_i, t) = \sigma \frac{\partial^2 y}{\partial t^2}\left(\frac{x_i + x_{i+1}}{2}, t\right).$$

What assumptions are inherent in this statement?

The limits of both expressions as $x_{i+1} \to x_i$ lead to a relation for each $x \in [0, L]$ and $t \geq 0$, namely,

$$T \frac{\partial^2 y}{\partial x^2} = \sigma \frac{\partial^2 y}{\partial t^2}.$$

This equation states that the vertical acceleration of any point on the string at time t is proportional to the curvature of the string at that point. For $y_{xx}(x, t)$ indicates how fast the slopes are changing in the vicinity of x at time t.

It is expected that the solution to this partial differential equation (known as the *one-dimensional wave equation*) will depend on the usual initial conditions. For example, the motion of a string and the properties of the resulting sound depend on the manner and point at which the string is bowed, hit, or plucked. These conditions are expressed in terms of the position and velocity of the string at a given time, usually initially. That is, the initial conditions are given by specifying the functions

$$y_0(x) = y(x, 0) \quad \text{and} \quad v_0(x) = \frac{\partial y}{\partial t}(x, 0) \quad \text{for} \quad x \in [0, L].$$

The fixed endpoints of the string give rise to other conditions called *boundary conditions*. Any solution function must satisfy, for all t, the equation

$$y(0, t) = 0 = y(L, t).$$

2.23 SEPARATION OF VARIABLES

The wave equation poses new problems. Any solution must be a function of two variables. The old techniques do not seem to apply. But this particular equation can be reduced to two ordinary differential equations. The procedure, called *separation of variables*, is a general method applicable to a significant class of partial differential equations. The solution is assumed to be of the form

$$y(x, t) = A(x)B(t),$$

and the functions A and B are determined by the differential equation and the boundary and initial conditions. For certain partial equations, $y(x, t) = A(x)B(t)$ is a solution if and only if A and B are solutions of certain ordinary equations. Of course, there may be other solutions of different types. This method only specifies solutions of the assumed form.

If $y(x, t) = A(x)B(t)$ is a solution of the wave equation, then

$$TA''(x)B(t) = \sigma A(x)\ddot{B}(t).$$

It is reasonable to assume that $A(x) \not\equiv 0$ on $[0, L]$ and $B(t) \not\equiv 0$ on $[0, \infty]$ because the trivial solution $y(x, t) = 0$ is of no interest. It simply states that no motion occurs. Thus, for $A(x) \neq 0 \neq B(t)$,

$$T \frac{A''(x)}{A(x)} = \sigma \frac{\ddot{B}(t)}{B(t)}.$$

This complete separation of variables in the differential equation has a useful implication. For any particular t, the expression on the right is constant. Therefore, if t is fixed even as x varies, the expression on the left must also be constant. That is, the expression $TA''(x)/A(x)$ is constant for $x \in [0, L]$. Similarly, a particular value of x determines a constant value for the left-hand expression. Thus, when x is fixed, all values of t must give a constant ratio $\sigma \ddot{B}(t)/B(t)$. The constant in both cases must be the same. Hence $y = A(x)B(t)$ is a solution of the partial differential equation if and only if A and B are solutions of the ordinary differential equations $TA''(x) = KA(x)$ and $\sigma \ddot{B}(t) = KB(t)$. Moreover, in this particular case, the ordinary equations are both linear with constant coefficients.

2.24 IMPLICATIONS OF THE BOUNDARY CONDITIONS

The boundary conditions may be stated as $A(0) = A(L) = 0$. Thus the constant K is unequal to 0. For if $K = 0$, then $A''(x) = 0$ and $A(x)$ is linear. But a linear function through $(0, 0)$ and $(L, 0)$ is the zero function. Similarly, $K \not> 0$ (see

Exercise 1 [2.24.1] below). Hence $K < 0$. It is convenient to write the ordinary equations as

$$\frac{T}{\sigma} A''(x) = -\omega^2 A(x) \quad \text{and} \quad \ddot{B}(t) = -\omega^2 B(t).$$

Then ω indicates the angular frequency of the function

$$B(t) = c_1 \cos \omega t + c_2 \sin \omega t.$$

The function

$$A(x) = c_3 \cos \sqrt{\sigma/T}\, \omega x + c_4 \sin \sqrt{\sigma/T}\, \omega x$$

must satisfy the boundary conditions. $A(0) = 0$ implies that $c_3 = 0$. $A(L) = 0$ implies that $c_4 \sin \sqrt{\sigma/T}\, \omega L = 0$. If $A(x) \neq 0$, then $c_4 \neq 0$ and $\sin \sqrt{\sigma/T}\, \omega L = 0$. Hence $\sqrt{\sigma/T}\, \omega L = n\pi$. Thus the boundary conditions are satisfied only for certain discrete values of ω, that is,

$$\omega = \frac{n\pi}{L} \sqrt{\frac{T}{\sigma}},$$

where n is any positive integer (only a positive frequency is meaningful). Hence the solution function is

$$y(x, t) = (\sin \sqrt{\sigma/T}\, \omega x)\,(c_1' \cos \omega t + c_2' \sin \omega t).$$

This is a quantum result. Each positive integer n determines a constant ω and a solution of the wave equation. The constant ω cannot change arbitrarily, but only in definite fixed jumps of $\pi/L \sqrt{T/\sigma}$ units.

The constants $c_i' = c_4 c_i$, $i = 1, 2$, are determined by the initial conditions. This appears to be a general solution, but it is only the general solution of the form $y(x, t) = A(x)B(t)$. There may be other solutions. For the vibrating string, this separable solution is a very useful model.

EXERCISES [2.24.1]

. . . .

1. Show that a positive constant K does not satisfy the boundary conditions for the string with fixed endpoints.

2. How does the period of the model of a string with uniformly distributed mass compare with the observed period, $2(mL/T)^{1/2}$ (see Section 2.21)?

3. Write the solutions for a plucked string that is released with velocity $v_0(x) = 0$ from the initial position $y_0 = y(x, 0)$.

4. Write the solution for a bowed string, satisfying the conditions $y_0(x) = y(x, 0) = 0$ and $v_0(x) = y_t(x, 0)$.

5. Discuss the effects of changes in the tension, length, or density of a string. Does this agree with your knowledge of the violin or piano? (A higher frequency has a higher pitch.)

6. Consider the frequencies and periods for different values of n. How are these related to the overtones of a musical note? Can you explain the technique of playing harmonics on the violin? (The octave above a given note is defined as the note with twice the frequency.)

7. State all the assumptions concerning the solution function that are necessary to derive the wave equation.

8. Show that the operator

$$\frac{\partial^2}{\partial t^2} - \frac{T}{\sigma} \frac{\partial^2}{\partial x^2}$$

is linear.

· ·

2.25 SUPERPOSITION, MUSICAL SOUNDS

Since the operator

$$\frac{\partial^2}{\partial t^2} - \frac{T}{\sigma} \frac{\partial^2}{\partial x^2}$$

is linear, the sum of any two solutions is a solution. Indeed, any linear combination of solutions is a solution. Even the series

$$\sum_{n=1}^{\infty} \sin \frac{n\pi}{L} x \left[a_n \cos \frac{n\pi}{L} \sqrt{\frac{T}{\sigma}} t + b_n \sin \frac{n\pi}{L} \sqrt{\frac{T}{\sigma}} t \right],$$

called a *Fourier series*, may be a solution.

Each value of n corresponds to a different mode of vibration (Fig. 2.18). For $n = 1$, $\sin (\pi/L)x = 0$ at $x = 0$ and L. And for each $x \in (0, L)$, the corresponding point oscillates with simple harmonic motion. The maximum amplitude occurs at $x = L/2$. For $n = 2$,

$$\sin \frac{2\pi}{L} x = 0 \qquad \text{at} \quad x = 0, \frac{L}{2}, L.$$

The string vibrates in two parts and the frequency of the motion is doubled. The note, by definition, is one octave above that for $n = 1$. For $n = 3$, the frequency is three times that of the fundamental note ($n = 1$). The pitch is in the second octave.

The western scale has 12 halftones in an octave. In the well-tempered scale, as used in modern piano tuning, successive tones are equally spaced. That is, the ratio of the frequencies of two successive notes is $2^{1/12}$. The second overtone

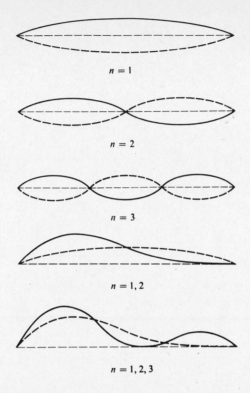

$n = 1$

$n = 2$

$n = 3$

$n = 1, 2$

$n = 1, 2, 3$ **Figure 2.18**

($n = 3$) is replaced by a note whose frequency is $2^{19/12}$ that of the fundamental ($3 = 2 \cdot \frac{3}{2} \doteq 2 \cdot 2^{7/12}$). The note is the fifth above the octave (count the halftones in twelfths), but the just tone and tempered tone differ. In the tempered scale, A$\sharp$ and B$\flat$ are the same and music in different keys can be played without retuning; not so in the just scale.

What is the physical interpretation of a solution that is the linear combination of solutions corresponding to different values of n? The different modes of vibration occur simultaneously; that is, the modes of vibration are superposed on one another. Thus a single string may produce the basic note ($n = 1$), and in addition, overtones corresponding to $n = 2, 3, 4, \ldots$ Hence a musical sound, as an impressed force, may activate several piano strings. The relative amplitudes of the various overtones determine the *quality* of the sound. Viola and oboe sounds differ in texture because of the differing strengths of the overtones.

Note that each solution, including the trigonometric or Fourier series, has a period of $2\sqrt{\sigma/T}\,L$. As Fourier indicated, almost all periodic functions may be expressed as a sum of sine and cosine functions. Since frequencies above a certain level are inaudible, the texture of music we hear is determined by a finite set of overtones. Mathematically and mechanically, a musical note may be broken down into its component parts or overtones, each of which is represented by a sinusoidal function. And the harmonic analysis may be reversed. Sinusoidal waves may be produced electronically and sounded simultaneously at chosen amplitudes to produce the sound of a particular musical instrument.

EXERCISES [2.25.1]

· ·

1. a) Determine the sequence of musical intervals for the overtones corresponding to $n = 2, 3, 4, 5, 6$.
   b) Compute the frequency ratio of the intervals in the tempered scale.
2. Sketch the function

$$f(x) = \sum_{n=1}^{3} a_n \sin \frac{n\pi}{L} x \qquad \text{on} \quad [0, L]$$

for different triplets (a_1, a_2, a_3). Choose the triplets to produce wide variations in the curves.

· ·

The harmonic oscillator has a single characteristic frequency. The vibrating string, a one-dimensional object, has many characteristic frequencies, one for each n, which produce harmonic overtones. A suspension bridge has the same boundary conditions as the taut string, and a corresponding set of characteristic frequencies— one for each integer.

A natural extension of the vibrating string is the vibrating membrane or drumhead. The two-dimensional vibration patterns are varied and striking. The characteristic frequencies, one for each pair of integers, are not necessarily harmonic (that is, they are not integral multiples of a common frequency). The circular membrane will be considered in Chapter 3 after series solutions and related numerical methods have been discussed.

The earth, a three-dimensional oscillator, has a characteristic frequency for each triplet of integers.

Other types of differential equations are solved by reduction to linear equations. Such an instance occurs in the demonstration that the inverse-square law of gravitation implies Kepler's first two laws.

2.26 THE LAW OF GRAVITATION IMPLIES KEPLER'S FIRST TWO LAWS

The gravitational law was derived from Kepler's laws in Section 1.15. Now the motion of a planet relative to the sun will be determined, assuming a gravitational force of magnitude $Gm_P m_S/r^2$ directed toward the sun (m_P and m_S indicate the masses of the planet and the sun, and r is the separation distance). Again, let S be the origin of the coordinate system and let $P = (x, y)$ represent the position of the planet. Newton's law $\mathbf{F} = m\mathbf{a}$, now a vector law, may be expressed in terms of the horizontal and vertical components (see Fig. 2.19):

$$m_p\ddot{x} = -G\,\frac{m_p m_s}{r^2}\cos\theta,$$

$$m_p\ddot{y} = -G\,\frac{m_p m_s}{r^2}\sin\theta.$$

$$(2.26.1)$$

The expectation of an elliptical orbit suggests initial conditions that locate the perihelion on the horizontal axis: $\mathbf{P_0} = (x_0, y_0)$ and $\mathbf{v_0} = (\dot{x}(0), \dot{y}(0)) = (0, v_0)$. The basic problem is to solve the differential equations and initial conditions and

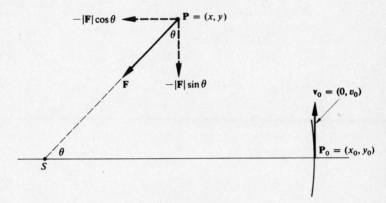

Figure 2.19

obtain an equation of motion. The procedure is to replace equations (2.26.1) by equivalent equations that yield to techniques already developed.

In order that the reader may carry through the argument, only methods and results are indicated as guidelines. The steps are listed as exercises and hints are given in the answer section.

EXERCISES [2.26.1]

1. *Equations (2.26.1) are equivalent to the system* $2\dot{r}\dot{\theta} + r\ddot{\theta} = 0$ *and* $\ddot{r} - r\dot{\theta}^2 = -Gm_s/r^2$.

   a) If x and y are expressed in terms of r and θ, the component equations become

$$(\ddot{r} - r\dot{\theta}^2)\cos\theta - (2\dot{r}\dot{\theta} + r\ddot{\theta})\sin\theta = -Gm_s\,\frac{\cos\theta}{r^2},$$

$$(\ddot{r} - r\dot{\theta}^2)\sin\theta + (2\dot{r}\dot{\theta} + r\ddot{\theta})\cos\theta = -Gm_s\,\frac{\sin\theta}{r^2}.$$

$$(2.26.2)$$

b) A simpler system may be obtained by multiplying equations (2.26.2) by suitable factors and combining results:

$$2\dot{r}\dot{\theta} + r\ddot{\theta} = 0,$$
$$\ddot{r} - r\dot{\theta}^2 = -Gm_S/r^2. \tag{2.26.3}$$

2. *The equation* $2\dot{r}\dot{\theta} + r\ddot{\theta} = 0$ *implies Kepler's second law.*

The use of an integrating factor (see Section 1.19) in the equation $2\dot{r}\dot{\theta} + r\ddot{\theta} = 0$ implies that $r^2\dot{\theta} = k$, a constant. But this is the mathematical statement of Kepler's second law (see Section 1.15).

3. *Equations* (2.26.3) *imply Kepler's first law.*

The initial conditions expressed in polar coordinates imply that $k = r_0 v_0$.

a) Hence $\dot{\theta} = r_0 v_0/r^2$ and $\dot{\theta}$ may be eliminated in equation $\ddot{r} - r\dot{\theta}^2 = -Gm_S/r^2$. The resulting second-order equation in r is not linear, but it may be reduced to a linear equation in $u(\theta)$ by letting $u(\theta) = 1/r$. This substitution, together with the relation $\dot{\theta} = r_0 v_0 u^2$, gives

$$u''(\theta) + u = Gm_S/r_0^2 v_0^2.$$

This particular substitution is hard to motivate. It is probably safe to say that many trial-and-error attempts preceded this ingenious reduction.

b) Show that the function satisfying this nonhomogeneous linear equation and the initial conditions may be written as

$$u = \frac{1}{r} = \left(\frac{1}{r_0} - \frac{Gm_S}{r_0^2 v_0^2}\right)\cos\theta + \frac{Gm_S}{r_0^2 v_0^2},$$

or

$$r = \frac{\dfrac{r_0^2 v_0^2}{Gm_S}}{1 + \left(\dfrac{r_0 v_0^2}{Gm_S} - 1\right)\cos\theta}.$$

. . . .

This last equation of motion is the polar equation of a conic section of eccentricity $e = r_0 v_0^2/Gm_S - 1$.

Observations of planets show that $v_0^2 < 2Gm_S/r_0$, so that $e < 1$, and the orbit of a planet is an ellipse.

Similar observations of comets provide a method of distinguishing between one-pass comets and those, like Halley's comet, that will return. v_0 and r_0 determine the eccentricity, which indicates whether or not the path is elliptical and periodic.

This same solution function serves as a model for the motion of satellites in the earth's gravitational field (ignoring the drag).

EXERCISES [2.26.2]

· ·

In a coordinate system centered at the earth, the orbit of a satellite may be described as

$$r = \frac{r_0^2 v_0^2 / G m_E}{1 + [(r_0 v_0^2 / G m_E) - 1] \cos \theta}.$$

1. Under what conditions will the orbit of a satellite be circular? (The circle, of course, may be considered as a special case of an ellipse.)

2. Kepler's second law, for a circular orbit, implies that the magnitude of the velocity is constant, that is $|\mathbf{v}| = v_0$. At the surface of the earth, $r = r_E$ and $G m_E / r_E^2 = g$. Using $r_E = 6400$ km (4000 mi), show that the speed $v = v_0$, as a function of the radius r_0 of a circular orbit, is $v = 6400/\sqrt{165 r_0}$ km/sec or $4000/\sqrt{165 r_0}$ mi/sec.

3. How long would it take a satellite to circle the earth "at the surface," assuming no atmosphere and hence no drag? How does this time compare with a round trip in a tunnel through the earth?

4. What must be the speed and period of a satellite circling the earth 1000 km above the surface?

5. The first sputnik had a nearly circular orbit and a period of 96 min. Estimate its height above the surface.

· ·

2.27 SOME REMARKS ON LINEAR DIFFERENTIAL EQUATIONS AND SERIES

The general solution of a second-order linear differential equation with constant coefficients is defined on R and can be written in terms of polynomial, exponential, and sinusoidal functions (Theorem 2.12.1). The solution forms categorize various types of oscillatory and/or damped motion. Higher-order equations with constant coefficients have similar solutions which may be obtained by similar methods.

But the restriction on the coefficients seems a stringent one. What can be said about linear equations with nonconstant coefficients? Existence Theorem 2.1.1 applies to linear equations whose coefficients are continuous on a common interval. Which of the subsequent theorems apply to this larger class of equations? How may solutions be obtained?

EXERCISES [2.27.1]

· ·

1. Which of Theorems 2.2.1 through 2.17.1 may be extended to the third-order equation $y''' + p_1 y'' + p_2 y' + p_3 y = r$ (or 0 as the case may be)? If uncertain, make a reasoned

guess. If possible, restate the theorem for the third-order equation. What complications arise as the order increases?* (*Note:* Theorem 2.1.1 does generalize to order 3. This is a consequence of the existence proof in Chapter 4.)

2. Conjecture solution forms for the third-order homogeneous linear equations with constant coefficients, for certain types of auxiliary equations. Discuss the algebraic problems that may arise as the degree of the auxiliary equation increases.*

3. What are the restrictions on p and q in each of Theorems 2.2.1 through 2.17.1? Which theorems apply to nonconstant coefficients?

. . . .

Theorem 2.5.1 holds for equations with nonconstant coefficients. A general solution of $y'' + p(x)y' + q(x)y = 0$ is determined by two independent solutions. And if p and q are well-behaved functions, there is a general procedure for obtaining specific solutions. The procedure is one already used, the method of undetermined coefficients (Section 2.17). The only difference here is the set of trial solutions; it is the set of functions defined by series of the form

$$\sum_{k=0}^{\infty} a_k x^k \quad \text{or} \quad \sum_{k=0}^{\infty} a_k (x - x_0)^k.$$

Many linear differential equations with nonconstant coefficients that serve as models in mathematical physics cannot be solved in terms of elementary functions. Series provide a rich source of new functions.

This second encounter with series solutions raises theoretical and pedagogical questions. How are operations performed on series? To be specific: which operations were used in Exercise 1 [2.17.1] in determining the coefficients of the trial solutions? Can these operations be applied in the same way to series? As you may suspect, the series above are useful just because, in many ways, they may be treated like polynomials.

A second question concerns the reader's familiarity with infinite series, which are often not included in first-year calculus. The reader can assume the operational properties of power series and use these properties to manipulate series and obtain solutions. Indeed, this is the way mathematicians proceeded in the eighteenth and nineteenth centuries. They developed techniques for well-behaved series and gradually increased the collection of controversial series that demanded more careful treatment.

But here, a basic understanding of series is needed for all the material that follows: series solutions of differential equations, numerical methods, Bessel functions, and even the existence and uniqueness theorem and its generalizations

---

*Theory and techniques for higher-order equations are found in Greenspan, Donald, *Theory and Solution of Ordinary Differential Equations.* New York: Macmillan, 1960, and Hochstadt, Harry, *Differential Equations.* New York: Holt, Rinehart & Winston, 1964.

developed in Chapter 4. Hence Chapter 3, which concerns series solutions and Bessel functions as models of the vibrating membrane, will start with an introduction to series.

CHAPTER 3

SERIES SOLUTIONS

3.1 INTRODUCTION

Chapters 1 and 2 have left unanswered questions and suggested new questions or extensions. What sort of existence theorem holds for the nonseparable equation $y' = f(x, y)$? How are solutions obtained for linear differential equations with nonconstant coefficients? What kinds of motion are possible for higher-dimensional oscillators, the natural extension of the harmonic oscillator, and the vibrating string? What are some of the more recent developments in numerical methods? How do the methods compare? These questions, which will be considered in Chapters 3 and 4, mark a turning point because they are best approached by methods involving sequences and series.

Since series are often not included in first year calculus, an introduction to sequences and series opens this chapter. The reader may skip this introduction (Sections 3.2 through 3.6), treat it as a review, or work it through carefully, depending on individual circumstances. Rigorous proofs are not presented. Instead, exercises are designed to develop an intuitive feeling for sequences, series, and the concepts used in subsequent sections. Indeed, since much of the work in this chapter is an extension of previous work, a significant part of the development is in the exercises, to be carried out by the reader.

In Chapter 3, series solution methods are presented and applied to linear differential equations with nonconstant coefficients. And series solutions are used to generate numerical approximations to a predetermined degree of accuracy. In particular, Bessel functions are introduced as solutions of the two-dimensional wave equation for the circular vibrating membrane. Numerical approximations are used to realize the physical implications. The Bessel equation also exhibits the complications that call for a more extensive theory of series solutions than is presented here.

An existence and uniqueness theorem for $y' = f(x, y)$ and a general discussion of numerical methods are given in Chapter 4.

3.2 NUMERICAL SEQUENCES

A numerical sequence $\langle a_n \rangle$ is an ordered set of real numbers, $a_1, a_2, \ldots, a_n, \ldots$. Each function f whose domain is the positive integers and whose range is in R defines a sequence $\langle a_n \rangle = \langle f(n) \rangle$. For example $f: n \to 1/2^n$ defines the sequence

$\langle 1/2^n \rangle = 1/2, 1/2^2, \ldots, 1/2^n, \ldots$ Since the set notation $\{\ \}$ indicates a disregard for order, the symbol $\langle a_n \rangle$ is used to denote the set $\{a_n\}$ together with the order imposed by the counting numbers.

The numbers $1/2^n$ seem to approach 0 as n increases. Sequences, whose terms approach a unique real number, are of particular interest. By definition, $\lim \langle a_n \rangle = L$, if and only if, for every number $\varepsilon > 0$, there exists a positive integer n_ε, such that for all $n > n_\varepsilon$, the difference $|a_n - L| < \varepsilon$. If $\lim \langle a_n \rangle = L$, then $\langle a_n \rangle$ is said to *converge to L*, the *limit of the sequence*. For example, consider the sequence $\langle 1/2^n \rangle$. Let ε be any positive number. It is always possible to find an integer k such that $\varepsilon > 10^{-k}$. And if $\varepsilon > 10^{-k}$, then for $n > 4k$, $|1/2^n - 0| < \varepsilon$. Hence $\lim \langle 1/2^n \rangle = 0$. Rephrased geometrically, $\lim \langle a_n \rangle = L$, if and only if, every ε-interval,

$$I = \{x : |x - L| < \varepsilon\},$$

contains all but a finite number of the a_n.

This definition replaces the intuitive idea of "approaches" or "gets closer and closer" with an arithmetical criterion. The definition is initially difficult to grasp. It has a further disadvantage. It offers no hint as to how to find L. The limit must be guessed before it can be tested. The number line may serve as an intuitive aid. If the numbers a_n are marked on a number line, do they appear to converge to a single point L?

EXERCISES [3.2.1]

1. Define distinct sequences $\langle a_n \rangle$ and $\langle b_n \rangle$ such that $\{a_n\} = \{b_n\}$.

   Exercises 2 through 7 apply to the sequences below.

   a) $\langle 1/2^n \rangle$ b) $\langle 2^n \rangle$ c) $\langle (-1)^n \rangle$
   d) $\langle n \rangle$ e) $\langle 1/n \rangle$ f) $\langle n/(n+1) \rangle$
   g) $\langle 1/n^2 \rangle$ h) $\langle \ln n \rangle$ i) $\langle 1 + (-1)^n \rangle$
   j) $\langle (-1)^n \, 1/n \rangle$ k) $\langle (-1)^n \, n/(n+1) \rangle$ l) $\langle (1 + 1/n)^n \rangle$

   m) $\langle 1/n! \rangle$ n) $\left\langle 1 + \dfrac{1}{1!} + \dfrac{1}{2!} + \cdots + \dfrac{1}{n!} \right\rangle$

   o) $\left\langle 4 \left(1 - \dfrac{1}{3} + \dfrac{1}{5} + \cdots + \dfrac{(-1)^{n-1}}{2n-1} \right) \right\rangle$

2. Write the first five terms of each sequence.

3. Indicate which sequences converge. In each case state the limit L and a suitable n_ε for $\varepsilon = 0.1$ and $\varepsilon = 10^{-10}$.

4. Which sequences diverge (do not converge)?

5. List the sequences which you cannot fit in either classification. Try to classify them later when more tests are available.

6. Which sequences are bounded, that is, for which sequences does there exist a number B such that $|a_n| \le B$ for all n? Indicate a bound B in each case.

7. Which sequences are nondecreasing, that is, $a_{n+1} \ge a_n$ for all n?

8. Give examples of sequences which are

   a) bounded and divergent,
   b) bounded, nondecreasing, and divergent,
   or conjecture a relevant theorem.

9. a) Can two different sequences have the same limit?
   b) Can a single sequence have two distinct limits?
   c) In each case, give an example or an argument to support your position.

. . . .

If it is difficult to determine the limit of a sequence, it is also discouraging to keep guessing if there is no certainty that a limit exists. Is it possible to determine whether a limit exists without using the limit? That is, criteria for convergence based on the sequence itself, and not the limit, are needed.

Property S.1

A nondecreasing bounded sequence of real numbers converges.

EXERCISES [3.2.2]

. . . .

1. Which sequences in Exercises [3.2.1] satisfy the hypothesis of S.1?

2. Are there sequences in Exercises [3.2.1] which are bounded and nonincreasing? List those which diverge and those which converge. State a Property S.1 concerning bounded, nonincreasing sequences.

3. Define a sequence which satisfies the hypothesis of S.1 and whose limit is not a rational number. Is there such a sequence in the last set of exercises?

4. a) Prove Property S.1. [*Hint:* Show that the least upper bound is the limit L. Assume that every bounded set of reals has a least upper bound.]
   b) Does Property S.1 hold on the set of rational numbers? Define a sequence $\langle a_n \rangle$ where each a_n is rational but $\lim \langle a_n \rangle = \pi$. Define $\langle a_n \rangle$ where each a_n is irrational and $\lim \langle a_n \rangle = \pi$. Define $\langle a_n \rangle$ where each a_n is irrational and $\lim \langle a_n \rangle = 0$.

*5. a) If $\lim \langle a_n \rangle = A$ and $\lim \langle b_n \rangle = B$, does $\lim \langle a_n + b_n \rangle = A + B$?
   b) Do $\langle -b_n \rangle$ and $\langle a_n - b_n \rangle$ converge?
   c) Use the property in part (a) to define two new convergent sequences based on those in Exercises [3.2.1].

6. One of today's unsolved problems concerns Euler's constant

$$\gamma = \lim \langle 1 + 1/2 + \cdots + 1/n - \ln n \rangle.^*$$

Show that the sequence is bounded and decreasing. Hence γ is well defined. But it is still not known whether γ is rational or irrational. [*Hint:* Determine whether each a_n and $a_{n+1} - a_n$ is positive or negative. The graph in Fig. 3.1 may be helpful.]

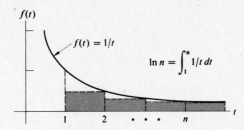

Figure 3.1

3.3 SEQUENCES OF FUNCTIONS

Consider a sequence of functions $\langle f_n \rangle$, where each f_n is defined on a common interval I. The sequence $\langle f_n \rangle$ *converges to* f_λ on I, if for each $x \in I$, the numerical sequence $\langle f_n(x) \rangle$ converges to $f_\lambda(x)$. That is, $\lim \langle f_n \rangle = f_\lambda$ on I, if and only if, for each $x \in I$ and each $\varepsilon > 0$, there is an integer $n_{x,\varepsilon}$ such that for all

$$n > n_{x,\varepsilon}|f_n(x) - f_\lambda(x)| < \varepsilon.$$

Note that a number $n_{x,\varepsilon}$ which suffices for a pair ε_1, x_1 might not suffice for the pair ε_1, x_2.

EXERCISES [3.3.1]

1. $\langle x^n \rangle$ 2. $\langle n^x \rangle$ 3. $\langle 1 + x^n \rangle$ 4. $\left\langle \dfrac{1}{1 + x^n} \right\rangle$ 5. $\left\langle \dfrac{x}{n} \right\rangle$

6. $\left\langle \dfrac{n}{x} \right\rangle$ 7. $\langle e^{nx} \rangle$ 8. $\langle (-x)^n \rangle$ 9. $\langle \sin nx \rangle$

For each sequence above, determine (if possible)

   a) two values of x for which the sequence converges,
   b) a value of x for which the sequence diverges,
   c) an interval on which the sequence converges.

*Courant, R., *Differential and Integral Calculus*, Vol. 1, 2nd ed., p. 381, New York: Interscience, 1937.

*10. For some of the sequences above, determine (if possible) an interval I on which the sequence converges, such that given $\varepsilon > 0$, a single n_ε suffices for every x in I. In this case the sequence is said to *converge uniformly* on I.

11. Consider the sequence $\langle f_n \rangle$ where $f_n(x) = x^n$ on R.

    a) Sketch curves for the functions $y = x^n$.

    b) What set of points appears to be the limit of these curves as n increases?

    c) Define $f_\lambda(x) = \lim \langle x^n \rangle$, for $x \in [0, 1]$.

    d) Are the f_n continuous? Is f_λ continuous on $[0, 1]$?

    e) Does $\langle f_n \rangle$ converge uniformly?

    f) Is the convergence uniform on $[0, \frac{1}{2}]$? That is, for $\varepsilon > 0$, can you determine N, such that $n > N \Rightarrow |0 - x^n| < \varepsilon$, for $x \in [0, \frac{1}{2}]$?

. . . .

3.4 NUMERICAL SERIES

The symbol $\sum_{k=1}^{n} a_k$ denotes the finite sum $a_1 + a_2 + \cdots + a_n$. The symbol

$$\sum_{k=1}^{\infty} a_k = a_1 + a_2 + \cdots + a_k + \cdots$$

has no meaning relative to the usual operation of addition. Another meaningless symbol is

$$
\begin{array}{ccc}
 & a_1 & \\
+ & & + \\
a_4 & & a_2 \\
+ & & + \\
 & a_3 &
\end{array}
$$

Addition is a binary operation that can be extended to any finite number of elements by applying the associative law. In both these symbols it is not clear when, if ever, the addition is completed. But $\sum_{k=1}^{\infty} a_k$ has intuitive meaning as a natural extension of the finite sums. For example, calculate $\sum_{k=1}^{n} 1/2^k$ for $n = 1, 2, 3, 4, 5, n$. Does the sequence $\langle \sum_{k=1}^{n} 1/2^k \rangle$ converge? The symbols $\sum_{k=1}^{\infty} 1/2^k$ and $\lim \langle \sum_{k=1}^{n} 1/2^k \rangle$ seem to have the same intuitive meaning. And the infinite sum is defined in terms of the finite sum:

$$\sum_{k=1}^{\infty} \frac{1}{2^k} = \lim \left\langle \sum_{k=1}^{n} \frac{1}{2^k} \right\rangle = 1.$$

In general, $\sum_{k=1}^{n} a_k$, called a *partial sum*, is denoted as S_n; and by definition

$$\sum_{k=1}^{\infty} a_k = \lim \langle S_n \rangle.$$

---

*Courant, R., *Differential and Integral Calculus*, Vol. 1, 2nd ed., p. 385ff, New York: Interscience, 1937.

$\sum_{k=1}^{\infty} a_k$ is said to *converge to S* if and only if $\lim \langle S_n \rangle = S$. S is called the sum of the series and denoted as

$$S = \sum_{k=1}^{\infty} a_k.$$

EXERCISES [3.4.1]

. . . .

1. Consider the series

   a) $\displaystyle\sum_{k=1}^{\infty} (-\tfrac{1}{2})^{k-1}$ 　　b) $\displaystyle\sum_{k=1}^{\infty} 10^{-k}$ 　　c) $\displaystyle\sum_{k=1}^{\infty} 3(10)^{-k}$

   d) $\displaystyle\sum_{k=1}^{\infty} 1/1^k$ 　　e) $\displaystyle\sum_{k=1}^{\infty} 1/k$ 　　f) $\displaystyle\sum_{k=1}^{\infty} (-1)^k$

   g) $\displaystyle\sum_{k=1}^{\infty} \frac{(-1)^{k-1}}{k}$ 　　h) $\displaystyle\sum_{k=1}^{\infty} 1/k!$ 　　i) $\displaystyle\sum_{k=1}^{\infty} 9(10)^{-k}$

   i) Write the first six terms, a_k, of each series. Determine the first six partial sums and S_n, if possible.
   ii) Which series converge and which diverge? Give the sum when possible. If you are uncertain, make a guess and classify the series as probably convergent or probably divergent. Support your guesses with intuitive arguments.

2. a) Write the series $1 + 1/2^3 + 1/3^3 + \cdots + 1/k^3 + \cdots$ using the summation symbol $\sum$ with three different sets of indices,

   $$\sum_{k=1}^{\infty} a_k, \quad \sum_{k=0}^{\infty} b_k, \quad \text{and} \quad \sum_{k=2}^{\infty} c_k.$$

   Check your notation by computing the first few terms in each case.
   b) Write the series $1 - 1/2^3 + 1/3^3 - \cdots + (-1)^{k+1} 1/k^3 + \cdots$ in the forms

   $$\sum_{k=1}^{\infty} a_k, \quad \sum_{k=2}^{\infty} b_k, \quad \text{and} \quad \sum_{k=-1}^{\infty} c_k.$$

   c) Write series (a), (b), (c), in Exercise 1, in the form $\sum_{k=0}^{\infty} a_k$. [*Note:* There are many symbols for a given series since the index may vary. But it is understood that the index number always increases by 1 in each successive term. It is often convenient to change indices when adding series.]

3. Consider the series $\sum_{k=0}^{\infty} a_k$ where each

   $$a_k = \frac{1}{2^k + 1} + \frac{1}{2^k + 2} + \cdots + \frac{1}{2^{k+1}}$$

   that is,

   $$a_0 = \tfrac{1}{2}, \quad a_1 = \tfrac{1}{3} + \tfrac{1}{4}, \quad a_2 = \tfrac{1}{5} + \tfrac{1}{6} + \tfrac{1}{7} + \tfrac{1}{8}.$$

a) Write numerical expressions for a_3, a_4, a_5.
b) Estimate S_n for $n = 0, 1, 2, 3, 4, 5$. Compare each a_n to $\frac{1}{2}$. Compare S_{n-1} to $n/2$. Does S_n converge?
c) What implications does this have for the *harmonic series* in Exercise 1(e) [3.4.1]?

4. A particularly useful series is the *geometric series* $\sum_{k=0}^{\infty} ar^k$; $a, r \in R$.

a) Which series in Exercise 1 are geometric?
b) Write geometric series for the number pairs (a, r) below.

$(1, \frac{1}{2})$, $(1, -\frac{1}{3})$, $(2, 2)$, $(100, \frac{1}{2})$, $(0.12, 0.01)$, $(1, 0)$, $(0, 1)$, $(0.1, 1)$, $(0.1, -1)$, $(0.01, 3)$, $(0.6, 0.1)$, $(5, 0.01)$

In each case, determine S_n, if possible. Which series converge (or probably converge) and which diverge?
c) Conjecture a theorem concerning the convergence of geometric series.
d) $S_n = \sum_{k=0}^{n} ar^k$ has $n + 1$ terms; $S_n - rS_n$ has only two. Calculate $S_n - rS_n$ and use it to express S_n in terms of a and r.
e) Determine $\lim \langle S_n \rangle$. Under what conditions does $\langle S_n \rangle$ converge?
f) Determine S for the geometric series in Exercises 1 and 4(b).

5. Consider the series

$$\sum_{k=1}^{\infty} \frac{1}{2^{k-1}} \quad \text{and} \quad \sum_{k=1}^{\infty} \frac{1}{k!}.$$

a) Compare the corresponding terms a_k of the two series.
b) Compare the partial sums S_n of the two series.
c) Discuss the convergence and sum of $\sum_{k=1}^{\infty} 1/2^{k-1}$.
d) Make a conjecture concerning the convergence of $\sum_{k=1}^{\infty} 1/k!$

....

3.5 CONVERGENCE TESTS

The ease of determining the convergence and sum of geometric series, as shown in Exercise 4 above, is a special situation. For nongeometric series, it is often difficult to determine sums. Hence Property S.1, applied to partial sums, is useful as an independent test for convergence. However, Exercise 5 above suggests a comparison test that can be applied directly to the series.

Property S.2

Let $\sum_{k=0}^{\infty} a_k$ be a series of nonnegative terms. Let $\sum_{k=0}^{\infty} b_k$ be a convergent series such that each $a_k \leq b_k$. Then $\sum_{k=0}^{\infty} a_k$ converges.

In order to apply this *comparison test*, one must first choose the convergent series $\sum_{k=0}^{\infty} b_k$. Hence it is helpful to collect convergent series.

EXERCISES [3.5.1]

. . . .

1. Show that the following series are convergent by comparing their terms to those of a suitable convergent series.

   a) $\sum_{k=0}^{\infty} \dfrac{1}{(1 + 2^k)}$ b) $\sum_{k=0}^{\infty} \left(\sin^2 \dfrac{1}{2^k}\right)\Big/ k!$ c) $\sum_{k=0}^{\infty} \dfrac{1}{2^k \, k!}$

   (It has proved convenient to define $0! = 1$. This is not surprising if you think of $k!$ as a multiplicative concept, for 1 is the multiplicative identity.)

2. Write a proof of Property S.2. [*Hint:* The two essential ideas are the comparison of the partial sums and Property S.1.]

3. Compare the series $\sum_{k=1}^{\infty} 1/2^k$ and $\sum_{k=1}^{\infty} 1/k!$. Can a comparison be made to show the convergence of $\sum_{k=1}^{\infty} 1/k!$? Are the early terms of a sequence or series important? Do the beginning terms affect the convergence or the sum? If $\sum_{k=6}^{\infty} a_k$ converges to S what can you say about the convergence and sum of $\sum_{k=1}^{\infty} a_k$?

4. Can the requirement in S.2 that each $a_k \le b_k$ be relaxed in any way? On the basis of Exercise 3, write a Property $\overline{\text{S.2}}$, similar to S.2, but with a slightly weaker hypothesis.

5. Conjecture a comparison test for divergence.

6. a) Let $\sum_{k=1}^{\infty} a_k$ and $\sum_{k=1}^{\infty} b_k$ be convergent series with sums A and B. Show that $\sum_{k=1}^{\infty} (a_k + b_k)$ converges to $A + B$. [*Hint:* See Exercise 5(a), [3.2.2]]
   b) If q is a constant and $\sum_{k=1}^{\infty} a_k = A$, show that $\sum_{k=1}^{\infty} q a_k = qA$.
   c) If $\sum_{k=1}^{\infty} a_k$ converges, does $\sum_{k=1}^{\infty} -a_k$?
   d) Given the hypothesis in (a) does $\sum_{k=1}^{\infty} (a_k - b_k)$ converge?

7. *The alternating series test* provides simple criteria for the convergence of series whose terms alternate in sign. Let $\langle a_n \rangle$ be a decreasing sequence, that is $a_n > a_{n+1}$ for all n, of positive terms such that $\lim \langle a_n \rangle = 0$. Then $\sum_{k=1}^{\infty} (-1)^{k+1} a_k$ converges.

   a) Plot the partial sums S_n on a number line and observe the implications of the conditions stated.
   b) Show that $\langle S_{2n} \rangle$ is increasing and bounded.
   c) State the corresponding properties for $\langle S_{2n+1} \rangle$.
   d) Determine $\lim \langle S_{2n+1} - S_{2n} \rangle$.
   e) What does the result in (d) imply for $\lim \langle S_{2n} \rangle$, $\lim \langle S_{2n+1} \rangle$, and $\lim \langle S_n \rangle$?
   f) Compare the values of

   $$|a_n| \qquad \text{and} \qquad \left| \sum_{k=n+1}^{\infty} (-1)^{k+1} a_k \right|.$$

   Discuss the error involved in using a finite sum to approximate the sum of an alternating series.
   g) Apply the test to the alternating series in Exercise 1, [3.4.1].
   h) Write a proof of the alternating series test.

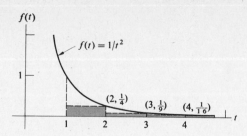

Figure 3.2

8. Sometimes series of integrals are used in the comparison test. Each term of the series $\sum_{k=2}^{\infty} 1/k^2$ may be interpreted as the area of a rectangle of base 1 and height $1/k^2$. (See Fig. 3.2.) Each area

$$\frac{1}{k^2} \leq \int_{k-1}^{k} \frac{1}{t^2}\, dt, \quad \text{and} \quad \sum_{k=2}^{\infty} \int_{k-1}^{k} \frac{1}{t^2}\, dt = \int_{1}^{\infty} \frac{1}{t^2}\, dt.$$

a) Show that each of the series

$$\sum_{k=2}^{\infty} \frac{1}{k^2}, \; \sum_{k=1}^{\infty} \frac{1}{k^2}, \; \sum_{k=1}^{\infty} \frac{1}{k^3}, \quad \text{and} \quad \sum_{k=1}^{\infty} \frac{1}{k^3 + 1}$$

converge.

b) Use integrals to show that the harmonic series $\sum_{k=1}^{\infty} 1/k$ diverges.

. . . .

The properties in Exercise 6 [3.5.1] above, may be rephrased in operational terms. The set of convergent numerical series is closed under the operation of addition. Indeed, two convergent series may be added by adding corresponding terms. The comparison test, the alternating series test, and the integral test (Exercises 4, 5, 7, 8 [3.5.1]) are each applicable to special types of series. The concept of absolute convergence leads to a more general test.

Property S.3

If a series $\sum_{k=1}^{\infty} |a_k|$ converges, then $\sum_{k=1}^{\infty} a_k$ converges and is said to *converge absolutely*. This implies that $\sum_{k=1}^{\infty} a_k$ converges for any arrangement of signs or terms. Thus the convergence of one series implies the convergence of a large class of related series. There are *conditionally convergent* series $\sum_{k=1}^{\infty} b_k$ that converge while $\sum_{k=1}^{\infty} |b_k|$ diverges. Such series have interesting properties. The sum and even the convergence can be changed by rearranging terms.

EXERCISES [3.5.2]

. . . .

1. Experiment with the series

$$\sum_{k=1}^{\infty} \frac{(-1)^{k-1}}{k} \quad \text{and} \quad \sum_{k=1}^{\infty} \frac{(-1)^{k-1}}{2^{k-1}}.$$

In each case, try to rearrange the order of the terms so that the partial sums are greater than or equal to the corresponding S_n in the original series. For the first series, determine a rearrangement with a greater sum.

2. Prove Property S.3. [*Hint:* Let S_n and $|S|_n$ denote the partial sums $\sum_{k=1}^{n} a_k$ and $\sum_{k=1}^{n} |a_k|$.] Let

$$S_n^+ = \sum_{\substack{k=1 \\ a_k > 0}}^{n} a_k \quad \text{and} \quad S_n^- = \sum_{\substack{k=1 \\ a_k < 0}}^{n} a_k.$$

   a) Use Property S.1 to show that $\langle S_n^+ \rangle$ and $\langle S_n^- \rangle$ converge if $\sum_{k=1}^{\infty} |a_k|$ converges.

   b) Express S_n in terms of S_n^+ and S_n^- and use this relation to show that $\langle S_n \rangle$ converges.

3. Show that if $\sum_{k=1}^{\infty} a_k$ converges, then $\lim \langle a_n \rangle = 0$.

4. Does the converse to the statement in Exercise 3 hold? Support your answer with an example or an argument. (See Exercises 3 [3.4.1], 8(b) [3.5.1].)

5. The geometric series suggests a test for absolute convergence; if a suitable ratio between successive terms is maintained, the series may be compared to a convergent geometric series.

   a) If

$$\left| \frac{a_{n+1}}{a_n} \right| \le q < 1 \quad \text{for all} \quad n \ge \text{some } n_0,$$

   show that $\sum_{k=1}^{\infty} |a_k|$ converges by comparison to

$$\sum_{k=0}^{\infty} |a_{n_0}| \, q^k.$$

   b) If

$$\lim \left\langle \left| \frac{a_{n+1}}{a_n} \right| \right\rangle = q < 1,$$

   show that $\sum_{k=1}^{\infty} a_k$ converges. [*Hint:* Can you find a number p, $q < p < 1$, such that for $n > n_0$,

$$\left| \frac{a_{n+1}}{a_n} \right| \le p?]$$

   c) Give examples of both convergent and divergent series satisfying

$$\lim \left\langle \left| \frac{a_{n+1}}{a_n} \right| \right\rangle = 1.$$

   d) If

$$\left| \frac{a_{n+1}}{a_n} \right| \ge q > 1, \quad \text{for} \quad n \ge n_0,$$

   show that $\sum_{k=1}^{\infty} a_k$ diverges.

   e) If

$$\lim \left\langle \left| \frac{a_{n+1}}{a_n} \right| \right\rangle = q > 1,$$

   does $\sum_{k=1}^{\infty} a_k$ diverge?

f) Summarize these properties in a statement relating the convergence of

$$\sum_{k=1}^{\infty} a_k \quad \text{and} \quad \lim \left\langle \left| \frac{a_{n+1}}{a_n} \right| \right\rangle .$$

g) Apply the *ratio test* as stated in (f) to the series in Exercise 1 [3.4.1].

h) Discuss the ratio test relative to absolute and conditional convergence.

6. a) List the convergence tests developed here for numerical series. Are they sufficient to decide the convergence of any series?

  b) Define a series whose convergence cannot be determined by any of the tests suggested so far.

. . . .

The *ratio test*, developed in Exercise 5 [3.5.2], is applicable to any series. In practice it is usually the first test applied. Comparison or alternating-series tests are used when $\lim \langle |a_{n+1}/a_n| \rangle = 1$. Stated formally, the *ratio test* asserts that if $\lim \langle |a_{n+1}/a_n| \rangle = q$, then $\sum_{k=0}^{\infty} a_k$

$$\begin{aligned} \text{converges absolutely} \quad &\text{if} \quad q < 1, \\ \text{diverges} \quad &\text{if} \quad q > 1, \\ \text{may converge or diverge} \quad &\text{if} \quad q = 1. \end{aligned}$$

There are series whose convergence is not determined by any of the tests given here. There are tomes of special tests for special series.

3.6 SERIES OF FUNCTION VALUES

Consider a series of function values $\sum_{k=0}^{\infty} f_k(x)$ where each f_k is defined on an interval I. $\sum_{k=0}^{\infty} f_k(x)$ *converges on* I, if and only if, for each $\bar{x} \in I$, the numerical series $\sum_{k=0}^{\infty} f_k(\bar{x})$ converges.

EXERCISES [3.6.1]

. . . .

1. State an ε, n definition for the convergence of $\sum_{k=0}^{\infty} f_k(x)$ to $f_\lambda(x)$ on I.

*2. Give an ε, n definition for the uniform convergence of $\sum_{k=0}^{\infty} f_k(x)$ on I.

3. Consider the series

a) $\displaystyle\sum_{k=0}^{\infty} 1/x^k$

b) $\displaystyle\sum_{k=0}^{\infty} x^k$

c) $\displaystyle\sum_{k=1}^{\infty} 1/kx$

d) $\displaystyle\sum_{k=0}^{\infty} (x - 2)^k/k!$

e) $\displaystyle\sum_{k=1}^{\infty} x/k^2$

f) $\displaystyle\sum_{k=0}^{\infty} e^{kx}$

g) $\displaystyle\sum_{k=1}^{\infty} x^k/k^2$ h) $\displaystyle\sum_{k=0}^{\infty} (\sin kx)/k!$ i) $\displaystyle\sum_{k=0}^{\infty} x^{2k}/2^k$

j) $\displaystyle\sum_{k=1}^{\infty} x^k/k$

For each series determine, if possible,

     i) two values of x for which the series converges,
     ii) a value of x for which the series diverges,
     iii) an interval on which the series converges,
     iv) an interval of absolute convergence,
*   v) an interval of uniform convergence.

4. a) Use the ratio test to show that the series $\sum_{k=0}^{\infty} a_k x^k$ converges absolutely for all x such that $|x| < \lim \langle |a_n/a_{n+1}| \rangle$.
  b) Discuss the convergence for x such that $|x| > \lim \langle |a_n/a_{n+1}| \rangle$. $\operatorname{Lim} \langle |a_n/a_{n+1}| \rangle = \rho$, is called the *radius of convergence*.
  c) Find the radius of convergence for the series $\sum_{k=0}^{\infty} a_k(x - x_0)^k$.

5. Consider the series $\sum_{k=0}^{\infty} x^k$ and $\sum_{k=0}^{\infty} (x - 1)^k$.

  a) Find the radii of convergence.
  b) Find the sum $S(x)$ of each of these geometric series. Compare $S(x)$ and the series for particular values of x.
  c) Sketch the graphs of $S_1(x) = \sum_{k=0}^{\infty} x^k$ and $S_2(x) = \sum_{k=0}^{\infty} (x - 1)^k$ on $(-1, 1)$ and $(0, 2)$ respectively. Discuss their similarities and differences.
  d) Does $\sum_{k=0}^{\infty} x^k$ converge at -1 and 1? Does $\sum_{k=0}^{\infty} (x - 1)^k$ converge at 0 and 2?
  e) Differentiate each series function term by term, as you would a polynomial, and find ρ for the resulting series. Compare the term-by-term differentiated series with $S_i'(x)$ for particular values of x, $x = 0.1, 0.2, 0.3, 0.4, 0.5$.

6. a) Differentiate the function
$$S(x) = \sum_{k=0}^{\infty} a_k x^k, \; x \in (-\rho, \rho),$$
term by term, as you would a polynomial. Compare the radii of convergence of the resulting and original series.
  b) Differentiate
$$S(x) = \sum_{k=1}^{\infty} \frac{\sin k^2 x}{k^2}$$
term by term. Compare the radii of convergence of the resulting and original series.

*7. Show that $\sum_{k=0}^{\infty} a_k x^k$ converges uniformly on any closed interval
$$I_r = \{x : |x| \leq r < \rho\}.$$

[*Hint:* Show that if $\sum_{k=0}^{\infty} a_k r^k$ converges absolutely, the terms $a_k r^k$ are bounded, that is, for some number M,
$$|a_k r^k| \leq M, \quad \text{for all } k.$$

Show that for $r \in (0, \rho)$, there is a number $\bar{r} \in (r, \rho)$ and $q \in (0, 1)$ such that $r = \bar{r}q$. Then

$$|x| \leq \bar{r}q.$$

Show that each $|a_k x^k| \leq Mq^k$, that is, $\sum_{k=0}^{\infty} |a_k x^k|$ may be compared to a convergent geometric series. Finally, given ε, there is an n_ε such that $n > n_\varepsilon$ implies that

$$M \left| \sum_{k=0}^{\infty} q^k - \sum_{k=0}^{n} q^k \right| < \varepsilon.$$

Show that the same n_ε applies to $\sum_{k=0}^{\infty} |a_k x^k|$, for all x in $[-r, r]$.]

. . . .

3.7 POWER SERIES

Exercises 4, 5, and 6 above suggest that series of the form $\sum_{k=0}^{\infty} a_k(x - x_0)^k$, called *power series*, have familiar and convenient properties. In some ways they act like finite sums or polynomials. Power series converge absolutely on the interval $(x_0 - \rho, x_0 + \rho) \equiv I_0$ and diverge for $|x - x_0| > \rho$, the radius of convergence. For each x in I_0, the series converges to a definite number, and hence defines a function. As suggested in Exercise 5 above, the functions defined by $\sum_{k=0}^{\infty} a_k x^k$ and $\sum_{k=0}^{\infty} a_k(x - x_0)^k$ differ only in their graph representations; one is a simple translation, $\bar{x} = x - x_0$, of the other. For simplicity, $\sum_{k=0}^{\infty} a_k x^k$ will be discussed; the particular uses of $\sum_{k=0}^{\infty} a_k(x - x_0)^k$ will be considered in the exercises.

Both $\sum_{k=0}^{\infty} a_k x^k$ and $\sum_{k=0}^{\infty} ka_k x^{k-1}$ converge uniformly on any closed interval in I_0, and this implies (see discussion below) that the series obtained by termwise differentiation does in fact converge to $f'(x) = (\sum_{k=0}^{\infty} a_k x^k)'$ for all $x \in I_0$. Indeed, since the derivative of a power series function is again a power series on the same interval, a power series on $(-\rho, \rho)$ has derivatives of all orders. [Exercise 6(b) [3.6.1] indicates that not all series have this property.] Since the interest here is in series as solution functions with derivatives on a common domain, only power series on intervals in $(-\rho, \rho)$ are considered. A series may converge at one or both endpoints, $\pm\rho$, but there is no guarantee that derivatives, even one-sided derivatives, exist at the endpoints. (See Exercise 1e, f [3.7.1].)

Since power series on I_0 may be differentiated as easily as polynomials, they provide a new class of solution functions. As trial solutions they may be substituted in linear differential equations and manipulated to determine coefficients. It is uniform convergence that insures this property of termwise differentiability. And uniform convergence implies other properties that are essential to the arguments in Chapter 4. Specifically, if a sequence of continuous functions converges uniformly on I, the limit function is continuous on I: moreover

$$\int_a^x \lim \langle f_n(u) \rangle \, du = \lim \left\langle \int_a^x f_n(u) \, du \right\rangle \qquad \text{for all} \quad a, x \in I.$$

These three implications of uniform convergence will be assumed and used, but not proved here. Such a theoretical diversion is inappropriate in an introductory differential equations book. But neither is it reasonable to ignore the property which might be characterized as the condition that ensures that limit functions are usable mathematical tools. Hence uniform convergence has been introduced and the theorems to be assumed have been stated. An excellent discussion of uniform convergence and proofs of its implications are given by Courant.* In this chapter, henceforth, power series will be differentiated term by term.

EXERCISES [3.7.1]

1. a) Which series in Exercise 3 [3.6.1] are power series?
   b) Determine the radii of convergence, ρ, of each power series.
   c) Write the derivative series.
   d) Which power series, and derivative series, converge on $(-\rho, \rho]$, $[-\rho, \rho)$, or $[-\rho, \rho]$?
   e) Show that $f(x) = \sum_{k=0}^{\infty} x^k/k^2$ converges on $[-1, 1]$, but $f'(1)$ and $f'(-1)$ are not defined.
   f) Indicate the series whose domain of convergence is not an interval.
   g) Consider $f(x) = (1 - x^2)^{1/2}$ on $[-1, 1]$. Is the derivative function defined at -1 or 1? Can one-sided derivatives be defined at the endpoints? (See Exercise 11(b), [3.8.1] for a series representation of this function.)

2. May power series be added term by term, that is, does the resulting series converge on the common interval of convergence? Does the same property hold for arbitrary series of functions? (See Exercise 5(a) [3.2.2].)

3. Consider $\sum_{k=0}^{\infty} a_k x^k + \sum_{k=0}^{\infty} k a_k x^{k-1}$ on I_0. Shift the index of one series so that the exponents agree (see Exercise 2 [3.4.1]) and add corresponding coefficients. Check the first few terms with the sum of the terms of the original series.

*4. Apply the property concerning integration of a uniformly convergent sequence of continuous functions, to the partial sum of an infinite series. State the corresponding theorem concerning integration of a uniformly convergent series. What is the common concept in the implications of uniform convergence for differentiation and integration of series?

What type of functions are defined by power series?

$$f_1(x) = \sum_{k=0}^{\infty} a_k x^k, \qquad a_k = \begin{cases} 1 & \text{if } k \leq 2 \\ 0 & \text{if } k \geq 3, \end{cases}$$

---

*Courant, R., *Differential and Integral Calculus*, Vol. 1, 2nd ed., p. 385ff, New York: Interscience, 1937.

defines the polynomial $f_1(x) = x^2 + x + 1$ on R. The expression

$$f_2(x) = \sum_{k=0}^{\infty} x^k$$

is an alternate definition of

$$f_2(x) = \frac{1}{1-x}$$

on $(-1, 1)$. (See Exercise 5 [3.6.1].) Familiar functions often need alternative definitions to provide numerical values. For example, how are trigonometric, exponential, and logarithmic functions evaluated? The statements $f_3(x) = e^x$ and $f_4(x) = \sin x$ provide no recipe for computing function values as does

$$f_1(x) = x^2 + x + 1.$$

Sine values may be defined as ratios of triangle segments, but angles and segments cannot be measured with the accuracy needed in sine tables. How, then, are function values obtained? These functions have power series definitions which provide numerical approximations to a predetermined degree of accuracy.

But in addition to defining functions already familiar to you, convergent power series offer a rich source of new functions. Many of these functions were first defined as solutions of linear equations with nonconstant coefficients. Here, Bessel functions will be introduced as solutions of a linear equation associated with a circular vibrating membrane. But first, series definitions and numerical approximations of familiar functions will be considered.

3.8 TAYLOR SERIES AND NUMERICAL METHODS

A series defining the familiar function $f(x) = e^x$ is obtained by assuming that

$$e^x = f(x) = \sum_{k=0}^{\infty} a_k x^k = a_0 + a_1 x + a_2 x^2 + \cdots + a_k x^k + \cdots \qquad \text{on } I, \text{ open,}$$

and determining coefficients that satisfy properties of the exponential function. The term a_0 may be evaluated by letting $x = 0$. And a_1 may be isolated, by differentiating the function, and then evaluated at $x = 0$.

$$e^0 = 1 = a_0.$$

$$(e^x)' = e^x = a_1 + 2a_2 x + \cdots + k a_k x^{k-1} + \cdots = \sum_{k=0}^{\infty} k a_k x^{k-1} \qquad \text{on } I.$$

$$e^0 = 1 = a_1.$$

$$(e^x)'' = e^x = 2a_2 + \cdots + k(k-1) a_k x^{k-2} + \cdots = \sum_{k=0}^{\infty} k(k-1) a_k x^{k-2} \qquad \text{on } I.$$

$$1 = 2a_2.$$

EXERCISES [3.8.1]

. . . .

1. a) Continue to differentiate and evaluate the series $e^x = \sum_{k=0}^{\infty} a_k x^k$ at $x = 0$. Show that $a_k = 1/k!$.
   b) Write the power series for e^x and determine ρ.

2. Are the functions $f_1(x) = e^x$ and $f_2(x) = \sum_{k=0}^{\infty} x^k/k!$ really equal? A series of operations does not necessarily produce a valid result. Do f_1 and f_2 both satisfy the differential equation $y' = y$ and a common initial condition? Use Theorem 1.13.1 to establish this equality. $\sum_{k=0}^{\infty} x^k/k!$ is called the *power series representation of* $f(x) = e^x$ at 0.

3. Sketch curves of $f(x) = e^x$ and the truncated series functions $f_n(x) = \sum_{k=0}^{n}(1/k!)x^k$ for $n = 0, 1, 2, 3, 4$ on $[-2, 2]$. Compare values of e^x and the truncated series at $x = 0, \frac{1}{2}, 1, 2$.

4. a) Estimate how many terms of the series are needed to evaluate $e^{1/2}$ and e^1 to three-place accuracy. How many terms are needed to compute tables on $[-\frac{1}{2}, \frac{1}{2}]$, or $[-1, 1]$, with three-place accuracy?
   b) Compute values for e^x on $[-\frac{1}{2}, \frac{1}{2}]$ to three significant decimals, using increments of 0.1.

5. Determine coefficients for the series

$$e^x = \sum_{k=0}^{\infty} a_k (x - \tfrac{1}{2})^k.$$

Use this *power series representation at* $\frac{1}{2}$ to compute values on $[\frac{1}{2}, 1]$. How many terms are needed for three-place accuracy here? Compare this to the estimate in Exercise 3.

6. Obtain a power series representation of $f(x) = \sin x$ at 0.

7. a) Carry through the same procedure for determining a_k for the function

$$f(x) = \sum_{k=0}^{\infty} a_k x^k.$$

Show that each $a_k = f^{(k)}(0)/k!$.

$$\sum_{k=0}^{\infty} \frac{f^{(k)}(0)}{k!} x^k$$

is called a Taylor series at 0. What assumptions must be made about f?
   b) Write a Taylor series at x_0 for $\sum_{k=0}^{\infty} a_k(x - x_0)^k = f(x)$. (Taylor series at $x_0 = 0$ are sometimes referred to as Maclaurin series.)

8. a) Determine Taylor series coefficients at 0 for the exponential, sine, and cosine functions. Do the derivatives of these Taylor series representations satisfy the usual relations?
   b) Sketch and compare curves for the sine function and its truncated Taylor series at 0 containing one, two, and three terms.

c) Discuss procedures and accuracy considerations in making tables for the sine and cosine functions. (See Exercise 7f [3.5.1].)

d) Which uniqueness theorem may be used to establish the equality of the sine and cosine functions and their Taylor series?

9. Use Taylor series to establish Euler's result,

$$e^{ibx} = \cos bx + i \sin bx,$$

as used in Section 2.12 (Treat i as a constant.)

10. Write a Taylor series for $\ln(1 + x)$ at 0, and for $\ln x$ at 1.

11. Write the first four terms of a Taylor series at 0 for

a) $f(x) = \tan x$ on $(-\pi/2, \pi/2)$.

b) $f(x) = (1 - x^2)^{1/2}$ on $[-1, 1]$. (See Exercise 1(g) [3.7.1].)

12. a) Show that each power series is the Taylor series of the function it defines. [*Hint:* Let $\sum_{k=0}^{\infty} a_k x^k = f(x)$ on $I_0 = (-\rho, \rho)$. Evaluate the function and its successive derivatives at 0 and determine the a_k.]

b) Show the same result for

$$\sum_{k=0}^{\infty} a_k(x - x_0)^k = f(x) \qquad \text{on} \quad I_0 = (x_0 - \rho, x_0 + \rho).$$

. . . .

Taylor series are power series by definition. Exercise 12 above shows that each power series on I_0 is also a Taylor series. In practice, it may be difficult to evaluate the higher derivatives (see Exercise 11 above). Functions not defined at 0 may have Taylor series at $x_0 \neq 0$. (See Exercise 10 above.)

3.9 TAYLOR POLYNOMIALS AND THE REMAINDER

A truncated power series,

$$\sum_{k=0}^{n} \frac{f^{(k)}(0)}{k!} x^k,$$

or a polynomial approximation, may be used to compute numerical values. (See Exercises 3, 4, 5, 8 [3.8.1].) Note that a truncated power series is a polynomial, but a polynomial approximation is not always a truncated power series. Thus a discussion of polynomial approximations will apply to truncated power series but will include polynomial approximations that are obtainable when power series are not. How accurate is an nth degree polynomial approximation $\sum_{k=0}^{n} a_k x^k$? Or, given a maximum allowable error, how many terms are needed to obtain this accuracy? The first question asks for a numerical evaluation of the difference

$$f(x) - \sum_{k=0}^{n} a_k x^k,$$

called the remainder and denoted $R_n(x)$. The second question asks for a number n_E such that the difference

$$\left| f(x) - \sum_{k=0}^{n_E} a_k x^k \right| = |R_{n_E}(x)| \leq E,$$

where E is the maximum error allowed. In both cases it is $R_n(x)$ that must be estimated. For a series, uniform convergence, by definition, ensures that a number n_E may be found for any closed interval in $(-\rho, \rho)$. At $x = 0$,

$$\sum_{k=0}^{n} \frac{f^{(k)}(0)}{k!} x^k = f(0) \qquad \text{and} \qquad R_n(0) = 0.$$

But $R_n(x)$ may increase as $|x|$ increases. (See Exercises 3, 4, and 8 [3.8.1].) As $|x|$ increases, the accuracy may be maintained by increasing the number of terms in the truncated series, or by using series $\sum_{k=0}^{n} a_k(x - x_0)$ over a succession of small intervals as $|x_0|$ increases. (See Exercise 5 [3.8.1].)

There are several neat expressions for $R_n(x)$; neat in that the form is very close to that of the $(n + 1)$-term of the series. The *Lagrange (1736–1813) form of the remainder* is

$$R_n(x) = f^{(n+1)}(\bar{x}) \frac{x^{n+1}}{(n + 1)!}$$

for some mean value $\bar{x} \in (-x, x)$, $x > 0$. A proof using the mean value theorem [actually the theorem of Rolle (1652–1716) is sufficient] is suggested in Exercise 5 below. Thus if there exists a bound M such that $|f^{(n+1)}(u)| \leq M$ for $u \in [-x, x]$, then

$$|R_n(x)| \leq M \frac{|x^{n+1}|}{(n + 1)!} \qquad \text{on} \quad [-x, x].$$

Thus $E = |R_n|$ may be estimated for a given n, or n_E may be determined to guarantee a certain maximum error E, on $[-x, x]$. For example, five terms of the series representation for e^x give function values with an error

$$E \leq e^{1/2} \frac{(\tfrac{1}{2})^5}{120} \leq 0.0004 \qquad \text{on} \quad [-\tfrac{1}{2}, \tfrac{1}{2}].$$

EXERCISES [3.9.1]

. . . .

1. Write the Lagrange form of the remainder for a Taylor polynomial at $x_0 \neq 0$. Determine a bound for $R_n(x)$ on $[x_0 - x, x_0 + x]$.

2. a) In approximating sine values with truncated Taylor series, compare estimates of the error E obtained by using the remainder $R_n(x)$ or the first unused term in the alternating series.

   b) Check the estimates made in Exercises 4 and 5 [3.8.1].

3. How many terms of the sine series are needed for a five-place table,

   a) using the series at 0 on $[0, \pi/2]$,

   b) using the series at 0 on $[0, \pi/4]$, and at $\pi/2$ on $[\pi/4, \pi/2]$?

4. Compare the number of terms needed for a three-place ln table on $[1, 1.5]$ and $[1, 2]$ using a Taylor polynomial for ln x at $x_0 = 1$.

5. [A proof to establish the Lagrange form of the remainder

$$R_n(x) = f(x) - \sum_{k=0}^{n} \frac{f^{(k)}(0)}{k!} x^k].$$

It is convenient to express $R_n(x)$ in a form close to the Taylor series term

$$\frac{f^{(k+1)}(0)}{(k+1)!} x^{k+1}.$$

To this end, let $R_n(x) = Ax^{n+1}$ and try to evaluate A in terms of the $(n+1)$-derivative. Assume that $n+1$ derivatives of f exist on an interval $[-b, b]$, where $b > 0$.

   Consider the difference

$$f(x) - \left(\sum_{k=0}^{n} \frac{f^{(k)}(0)}{k!} x^k + Ax^{n+1} \right) = F(x).$$

a) Is F a differentiable function? How many derivatives exist and on what interval?

b) Show that $F^{(k)}(0) = 0$ for $k = 0, 1, \ldots, n$.

c) Evaluate A, so that $F(b) = 0$. [Let A be that number for which $F(b) = 0$.]

d) Use Rolle's theorem to show that there exist numbers

$$c_1 \in (0, b) \quad \text{such that} \quad F'(c_1) = 0,$$
$$c_2 \in (0, c_1) \quad \text{such that} \quad F''(c_2) = 0,$$
$$c_n \in (0, c_{n-1}) \quad \text{such that} \quad F^{(n)}(c_n) = 0.$$

   In each case, show that the hypothesis of Rolle's theorem holds.

e) Is there a number $c_{n+1} \in (0, c_n)$ such that

$$F^{(n+1)}(c_{n+1}) = 0?$$

   What property must $f^{(n+1)}$ satisfy to ensure that c_{n+1} exists?

f) Evaluate $F^{(n+1)}(c_{n+1})$ and show that if $f^{(n+1)}$ is continuous on $(0, b)$,

$$A = \frac{R_n(b)}{b^{n+1}} = \frac{f^{(n+1)}(c_{n+1})}{(n+1)!}.$$

---

*See Wolfe, James, *Am. Math. Month.* **60**, 415, 1935.

That is, for some mean value $\bar{x} \in (0, b)$,

$$R_n(b) = \frac{f^{(n+1)}(\bar{x})}{(n + 1)!} b^{n+1}.$$

And for any x in $[0, b]$,

$$R_n(x) = \frac{f^{(n+1)}(\bar{x})}{(n + 1)!} x^{n+1},$$

where $\bar{x} \in (0, x)$.

g) Write a theorem concerning a polynomial representation for $f(x)$ on the interval $[0, b]$.

h) Does the same argument apply to the interval $[-b, 0]$?

i) Compare the theorem for $n = 1$ to the mean value theorem.

j) Write a similar theorem for a polynomial approximation

$$\sum_{k=0}^{n} \frac{f^{(k)}(0)}{k!} (x - x_0)^k.$$

. . . .

3.10 SERIES SOLUTIONS OF DIFFERENTIAL EQUATIONS

A Taylor series solution may sometimes be determined directly from a differential equation. For example, if $y' = y$ and $y(0) = 1$, then by induction every $y^{(k)}(0) = 1$; the solution has the series representation $\sum_{k=0}^{\infty} (1/k!)x^k$. The more general method of undetermined coefficients is applicable to a large class of linear equations. It is essentially the procedure outlined in Section 2.17, but the trial solutions are power series.

EXERCISES [3.10.1]

. . . .

1. List the operations used in determining the coefficients of the trial polynomial solutions in Exercise 1(b) [2.17.1]. Which operations have been validated for power series? Which have not?

2. Apply the same method to determine the a_k if

$$f(x) = \sum_{k=0}^{\infty} a_k x^k$$

on I is a trial solution of the differential equation and initial condition $y' = y$; $y(0) = 1$.

a) Substitute $y = \sum_{k=0}^{\infty} a_k x^k$ and its derivative in the differential equation.

b) Shift the index of one series so that the exponents agree, and add corresponding coefficients. (See Exercises 3 and 4 [3.7.1].)

c) The power series in the resulting equation,

$$\sum_{k=1}^{\infty} (ka_k - a_{k-1})x^{k-1} = 0,$$

defines the zero function. Hence each coefficient must be zero (see discussion below) and $a_k = a_{k-1}/k$ for $k \geq 1$.

d) Evaluate each a_k in terms of a_0. Use the initial condition to evaluate a_0. Write the series solution. Determine p. Does this agree with previous solutions of these equations?

. . . .

All the operations in Exercise 2 have been discussed and justified except those in 2(c). Is the series $\sum_{k=0}^{\infty} a_k x^k$, with all $a_k = 0$, the unique power series representation of the zero function on I? It is just as easy to answer a more general question. Can two distinct series $\sum_{k=0}^{\infty} a_k x^k$ and $\sum_{k=0}^{\infty} b_k x^k$ represent the same function f on $I \supset \{0\}$? Suppose

$$\sum_{k=0}^{\infty} a_k x^k = f(x) = \sum_{k=0}^{\infty} b_k x^k \qquad \text{on} \quad I \supset \{0\}.$$

Let $g(x) = f(x) - f(x) = 0 = \sum_{k=0}^{\infty} (a_k - b_k)x^k$ on I. All derivatives of g on I are zero functions. The coefficients $(a_k - b_k)$ may be determined, as before, by evaluating

$$g(x) = \sum_{k=0}^{\infty} (a_k - b_k)x^k$$

and its derivatives at $x = 0$. (See Section 3.8.) Then $k!(a_k - b_k) = 0$ and $a_k = b_k$ for $k \geq 0$. The two series, then, are not distinct on I.

The equation $a_k = a_{k-1}/k$, called a *recurrence relation*, defines a_k in terms of the previous coefficients $a_j, j \leq k$. (See Exercise 2(c) [3.10.1].) Such a relation may be applied over and over again until each a_k is expressed in terms of the early coefficients. If there is a recognizable pattern, the general coefficient a_k may be conjectured and established by induction. This whole process is possible because differentiation and addition of power series are term-by-term operations and may be applied directly to the coefficients and powers in each term. Rephrased, a linear differential equation imposes a recurrence relation on the coefficients of a power series function.

EXERCISES [3.10.2]

. . . .

1. Determine a series solution $\sum_{k=0}^{\infty} a_k x^k$ of the equation $y'' + y = 0$ with
   a) initial conditions $y(0) = 0$, $y'(0) = 1$,
   b) initial conditions $y(0) = 1$, $y'(0) = 0$,

c) no initial conditions.

Determine the radius of convergence in each case. (Note that without initial conditions, a_0 and a_1 are arbitrary and play the role of the arbitrary constants in the general solution.)

2. Determine a series solution for $y'' - xy = 0$. This equation has nonconstant co-efficients. What other property of series, already established, must be used in this case? (See Exercise 6(b) [3.5.1].) Obtain a general solution and verify that it is such.

3. Determine a solution for $y'' - 2xy' + 2y = 0$. Obtain a general solution and verify that it is such.

. . . .

   The series solutions in Exercises 2 and 3 are unrecognizable and seem to define new functions. Many differential equations have series solutions which cannot be identified in terms of elementary functions. The second-order linear equations with nonconstant coefficients include among their solutions many of the special functions of applied mathematics. These equations and their well-tabulated solution functions have taken the names of the men who investigated them. The Bessel equation is considered here as a model of the circular vibrating membrane, a two-dimensional analog of the vibrating string.

3.11 THE VIBRATING MEMBRANE

Consider a taut membrane fastened securely about a ring of radius r_0. On a drum (Fig. 3.3), or tambourine, this membrane is usually a disk of skin. If the membrane

Figure 3.3

is struck and then released, what is the resulting motion? It can be observed that the membrane vibrates with amplitudes that are small relative to the diameter of the disk. The taut membrane, like the stretched string, breaks under large displacements.

   The simplifying assumptions used to obtain a simple model are essentially the same as those for the string. Each point on the membrane is assumed to move on a line perpendicular to the equilibrium position of the membrane. The only force considered is that due to the tension in the membrane. At any point there is an outward force along every direction in the tangent plane. Consider instead, tension acting on a small segment. It is assumed that two equal and opposite forces are directed away from the midpoint, perpendicular to the segment, and in the tangent plane at the point (Fig. 3.4). The magnitude is proportional only to the measure of the segment, that is, the tension T per unit length is assumed to be constant both in position and time.

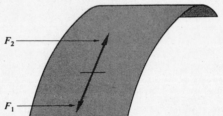

Figure 3.4

A distinguishing feature of this membrane is that its position is fixed on a circular boundary. Rectangular or elliptical boundaries give rise to quite different sorts of vibration patterns. This difference was demonstrated as early as 1787 by Chladni, who sprinkled sand on vibrating plates and watched the sand gather on the nodal curves, that is, points of no vibration. The curve in the boundary condition suggests the type of coordinates. In this case, the circle is most easily represented in polar coordinates with the pole at its center. If the equilibrium position of the membrane is in a horizontal plane, the height of each point at any time t is represented by a single number z. The motion is then described by a function $z = z(r, \theta, t)$ which is assumed to have continuous first and second partials. Polar coordinates, here, refer to the space domain of the function $z = f(r, \theta, t)$, just as the resulting equation is known as the two-dimensional wave equation.

The membrane is treated as a two-dimensional surface; thickness is recognized only in the assumption of a uniform density of σ units per square unit of area. Thus the mass is assumed to be uniformly distributed throughout the membrane. The motion at a point will be obtained as a limiting situation of the motion of a small piece of membrane. In polar coordinates, the simplest section is one bounded by small increments in r and arc lengths $r_i\theta$, that is, a truncated sector (Fig. 3.5) determined by (r_1, θ_1), (r_1, θ_2), (r_2, θ_1), and (r_2, θ_2). The mass of this sector is assumed to be constant throughout the motion. Four external forces are considered to be acting on this section, each directed perpendicularly away from the midpoint of an edge. The magnitudes of the forces are

$$|F_1| = T r_1 (\theta_2 - \theta_1), \qquad |F_3| = T (r_2 - r_1),$$
$$|F_2| = T r_2 (\theta_2 - \theta_1), \qquad |F_4| = T (r_2 - r_1).$$

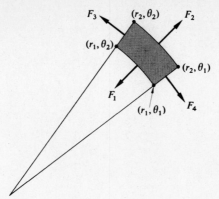

Figure 3.5

The vertical component of each force is $|F_i| \sin \alpha_i$, where α_i is the angle from the horizontal to the force vector (Fig. 3.6). As before (see Section 2.22), the vertical

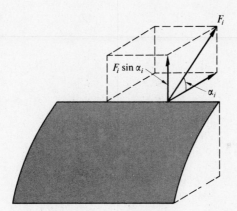

Figure 3.6

components may be written as derivatives, since $|F_i| \sin \alpha_i \doteq |F_i| \tan \alpha_i$ for small α_i. Tan α_i may be expressed as a derivative representing the slope in the direction of the horizontal projection of F_i. Hence

$$\tan \alpha_i = \frac{\partial z}{\partial r} (r_i, \tfrac{1}{2}(\theta_1 + \theta_2), t), \qquad i = 1, 2.$$

But tan α_i, $i = 3, 4$, is more complicated. A slope is the ratio of two linear measures, while z_θ involves the ratio of linear and angular measures. Tan α_3 may be determined as a derivative along the arc $r = \tfrac{1}{2}(r_1 + r_2)$ (see Fig. 3.7). Since the distance

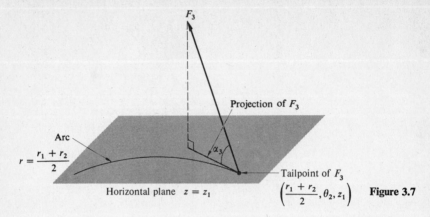

Figure 3.7

between two points (r, θ) and (r, θ_2) on the arc is $\frac{1}{2}(r_1 + r_2)(\theta - \theta_2)$, $\tan \alpha_3$ at $(\frac{1}{2}(r_1 + r_2), \theta_2, t)$ is

$$\lim_{\theta \to \theta_2} \frac{z(\frac{1}{2}(r_1 + r_2), \theta, t) - z(\frac{1}{2}(r_1 + r_2), \theta_2, t)}{\frac{1}{2}(r_1 + r_2)(\theta - \theta_2)}$$

$$= \frac{1}{\frac{1}{2}(r_1 + r_2)} \frac{\partial}{\partial \theta} z(\frac{1}{2}(r_1 + r_2), \theta_2, t).$$

EXERCISES [3.11.1]

. . . .

1. Give physical interpretations of z_t, z_{tt}, z_θ, z_r, z_{rr}, where

$$z_t = \frac{\partial z}{\partial t} \quad \text{and} \quad z_{tt} = \frac{\partial^2 z}{\partial t^2}.$$

2. Write the vertical components $|F_i| \sin \alpha_i$ in terms of r_i, θ_i, z_θ, and z_r. (As before, the angles at opposite edges vary by approximately π; opposite forces have vertical components of opposite sign.)

3. Write $|F_3| \tan \alpha_3 + |F_4| \tan \alpha_4$ as a single term, involving $z_{\theta\theta}$. (See Section 2.22). List all the necessary assumptions.

4. Express the vertical component of $F_1 + F_2$ in terms of z_{rr} and z_r. List the necessary assumptions. [*Hint:* The $|F_i| \tan \alpha_i$, $i = 1, 2$, contain different r_i. Use the relation $r_2 = r_1 + (r_2 - r_1)$ to break $|F_2| \tan \alpha_2$ into two terms. One term combines with $|F_1| \tan \alpha_1$ (using the mean value theorem). The term with coefficient $(r_2 - r_1)$ remains as a first derivative. Another approach is to consider $|F_i| \tan \alpha_i$, $i = 1, 2$, as functions of the form

$$f(r) = Kr \frac{\partial}{\partial r} z \left(r, \frac{\theta_1 + \theta_2}{2}, t \right)$$

and apply the mean value theorem to this function.]

5. Show that the vertical component of the total force is

$$T(\theta_2 - \theta_1)(r_2 - r_1)\left[r_1 z_{rr}\left(\bar{r}, \frac{\theta_1 + \theta_2}{2}, t\right) + z_r\left(r_2, \frac{\theta_1 + \theta_2}{2}, t\right)\right.$$

$$\left. + \frac{2}{r_1 + r_2} z_{\theta\theta}\left(\frac{r_1 + r_2}{2}, \bar{\theta}, t\right)\right].$$

6. The mass of the sector is the product of the area and the density per square unit. Show that at the centroid (r_c, θ_c) of the sector,

$$ma = \sigma\tfrac{1}{2}(r_1 + r_2)(r_2 - r_1)(\theta_2 - \theta_1) z_{tt}(r_c, \theta_c, t).$$

7. Assume that F acts at (r_c, θ_c).

   a) Write Newton's law in terms of r_1, θ_i, z and its partials.
   b) Simplify the equation and consider the limit as $r_2 \to r_1$ and $\theta_2 \to \theta_1$. The equation may be multiplied by $1/(r_2 - r_1)(\theta_2 - \theta_1)$ since neither difference is zero.
   c) Show that the partial differential equation

$$\frac{\sigma}{T} z_{tt} = z_{rr} + \frac{1}{r} z_r + \frac{1}{r^2} z_{\theta\theta}$$

   holds for each (r, θ, t) where $|r| < r_0$.

8. Show that the partial differential equation for a rectangular membrane is

$$\frac{\sigma}{T} z_{tt} = z_{xx} + z_{yy}.$$

   [*Hint:* Use rectangular coordinates and a rectangular section, and follow the procedure outlined in the exercises above.]

9. Show that the wave equation for the circular membrane may be obtained from the equation for the rectangular membrane, by the change of variables, $x = r \cos \theta$ and $y = r \sin \theta$. [Note that both differential equations in Exercises 7(c) and 8 are special cases of the wave equation

$$\frac{\sigma}{T} z_{tt} = \nabla^2 z.$$

   The Laplacian operator ∇^2, which indicates the effect of certain forces acting at a point of a surface, takes on different forms under different coordinate systems.]

. . . .

In the two-dimensional wave equation first published by Euler in 1764,

$$\frac{\sigma}{T} z_{tt} = z_{rr} + \frac{1}{r} z_r + \frac{1}{r^2} z_{\theta\theta}, \tag{3.11.1}$$

the function $z = f(r, \theta, t)$ is on a three-dimensional domain. Even so, separation of variables can still be carried out, this time in several steps. A procedure for this is suggested in Exercises 5 [3.12.1] and 3 [3.15.1].

3.12 NORMAL SYMMETRIC MODES OF VIBRATION

Another approach to the two-dimensional wave equation is to simplify the problem by restricting the set of solutions. For example, consider only solutions representing *normal modes of vibration* in which each point enjoys simple harmonic motion with a common period and phase. Then solutions have the form

$$z = f(r, \theta)A \cos(\omega t + \beta), \qquad \omega > 0$$

since ω indicates a frequency. If the solutions are further restricted to those with circular symmetry, $f(r, \theta)$ is independent of θ. Then the solutions

$$z = f(r)A \cos(\omega t + \beta)$$

represent *normal symmetric modes* of vibration. If the membrane is initially at rest in a displaced position, the initial conditions are $z_t(r, \theta, 0) = 0$ and $z(r, \theta, 0) = f(r)$.

EXERCISES [3.12.1]

1. Show that a solution representing a normal symmetric mode and zero initial velocity has the form $z = f(r) \cos \omega t$. [The constant may be absorbed in $f(r)$.]

2. Show that for solutions of the form $z = f(r) \cos \omega t$, the two-dimensional wave equation in polar coordinates implies that $f(r)$ is a solution of the ordinary equation

$$f''(r) + \frac{1}{r} f'(r) + \omega^2 \frac{\sigma}{T} f(r) = 0. \tag{3.12.1}$$

3. a) Denote the constant $\omega \sqrt{\sigma/T}$ by Q and show that the usual series methods applied to Eq. (3.12.1) yield solutions of the form

$$f(r) = \sum_{k=0}^{\infty} a_k r^k = a_0 \sum_{k=0}^{\infty} (-1)^k \frac{(Qr)^{2k}}{(k! \, 2^k)^2}. \tag{3.12.2}$$

   b) Find the radius of convergence.

4. a) Are any two of the solution series in Exercise 3, for different values of a_0, linearly independent? Are the coefficient functions p and q in Eq. (3.12.1) continuous?

   b) Show that for a circular membrane, struck at the center point, the initial displacement of this point determines the solution representing normal symmetric modes of vibration. (Assume that the membrane has initial zero velocity when released.) That is, show that the initial condition $z(r, \theta, 0) = f(r)$, applied at $r = 0$, yields the solution

$$f(r) = f(0) \sum_{k=0}^{\infty} (-1)^k \frac{(Qr)^{2k}}{(k! \, 2^k)^2}.$$

5. *Separation of variables.* Assume that

$$z(r, \theta, t) = f_1(r)f_2(\theta)f_3(t).$$

Substitute this solution in the partial differential equation and isolate f_3 and its derivatives. Show that $f_3''(t) - cf_3(t) = 0$. [*Hint:* Use arguments similar to those in Section 2.23]. As before, it may be shown that the boundary and initial conditions imply that $c < 0$. Assume this, and discuss the implications for the solution function and the motion. What is the implication of zero initial velocity on the solution function?

. . . .

The coefficient $p(r) = 1/r$ in the second-order linear equation (3.12.1),

$$f''(r) + \frac{1}{r}f'(r) + \omega^2 \frac{\sigma}{T} f(r) = 0,$$

is not continuous at 0. And yet the series method yields a solution on R, which includes the point of discontinuity. But Exercise 4 suggests that a general solution may not be obtainable by the usual procedure. For certain types of discontinuous functions p and q, particular series solutions may be found by the usual method. But more complicated methods may be necessary to obtain general solutions. For Eq. (3.12.1), a general solution contains the function (3.12.2) and a second solution of the form $\ln (\omega\sqrt{\sigma/T}\, r)$, which is not finite at $r = 0$. The finiteness of z for $r \leq r_0$ is a type of boundary condition which implies that in a general solution, the coefficient of an unbounded term on $[0, r_0]$ must be zero. Thus the solution already obtained,

$$f(r) = f(0) \sum_{k=0}^{\infty} (-1)^k \frac{(Qr)^{2k}}{(k!\, 2^k)^2},$$

is an appropriate model for normal symmetric modes of vibration.

3.13 THE BESSEL FUNCTION J_0 AND NUMERICAL APPROXIMATIONS

The solution function (3.12.2) with unit constant is called the *Bessel function of order 0* and is denoted as J_0:

$$J_0(x) = \sum_{k=0}^{\infty} (-1)^k \frac{x^{2k}}{(k!\, 2^k)^2}, \qquad x \in R,$$

J_0 cannot be expressed simply in terms of elementary functions, but a numerical approximation for J_0 on the interval $[0, 9]$ indicates the physical implications of the solution.

EXERCISES [3.13.1]

. . . .

1. Use truncated series to construct a numerical table with three-place accuracy for J_0 on the interval $[0, 9]$, using steps of 0.2. [Note that J_0 is an alternating series. If a computer is not used, construct a table on $[0, 2]$ and obtain the remaining values from the table in Appendix C.]

2. Sketch the numerical approximation constructed in Exercise 1. Is J_0 periodic?

3. Estimate and list the zeros α_i of J_0, that is, the set $\{x: J_0(x) = 0\}$ in the interval $[0, 9]$.

*4. Discuss the difficulties in specifying a series for J_0 at $x = 2$.

. . . .

3.14 THE BOUNDARY CONDITION $z(r_0, \theta, t) = 0$

The basic property of the circular membrane is that every point on the circle $r = r_0$ maintains a constant position throughout the motion: $z(r_0, \theta, t) = 0$. The circular boundary suggests the coordinate system, and hence the form of the wave equation, but it has not yet been imposed on the solution. For the simple harmonic solutions with circular symmetry, namely

$$z = f(0)\, J_0(Qr)\, \cos\, \omega t,$$

the boundary condition implies that $J_0(\omega\sqrt{\sigma/T}\, r_0) = 0$. Hence $\omega\sqrt{\sigma/T}\, r_0$ must be a zero α_i of J_0. The values σ, T, and r_0 are fixed for a given membrane. Thus it is the frequency ω that is affected by the boundary condition; ω may not assume arbitrary values, but only the values

$$\omega = \sqrt{\frac{T}{\sigma}}\, \frac{\alpha_i}{r_0}.$$

Thus the boundary condition imposes a quantum effect here, just as it did for the vibrating string. For a string, each frequency is an integer multiple of the fundamental frequency; the jumps in successive values of ω are constant. But for the drumhead, the frequency jumps are irregular, because the zeros α_i are not spaced at regular intervals. Actually the sequence $\langle \alpha_i - \alpha_{i-1} \rangle$ is increasing and approaches π. It has recently been shown that all the zeros of J_0 are irrational.

The lowest possible, or fundamental, frequency is

$$\omega = \sqrt{\frac{T}{\sigma}}\, \frac{\alpha_1}{r_0} \doteq \frac{2.405}{r_0}\, \sqrt{\frac{T}{\sigma}}.$$

The corresponding solution is

$$z = f(0)J_0\left(\alpha_1 \frac{r}{r_0}\right) \cos\, \omega t, \qquad r \leq r_0.$$

$f(0)$ is determined by the initial displacement. As r varies from 0 to r_0, the argument $\alpha_1(r/r_0)$ varies from 0 to α_1. Thus, at any time t, the curve of each radial section of the membrane is a multiple of the J_0 curve on $[0, \alpha_1]$, compressed or stretched onto the interval $[0, r_0]$ (see Fig. 3.8). The corresponding motion is called

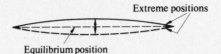

Extreme positions

Equilibrium position   **Figure 3.8**

the *first normal symmetric mode of vibration*. This is the simple vibration that one commonly visualizes. Seen across a section through a diameter, the membrane vibrates between the two extreme positions.

The frequency of the second mode of vibration is

$$\omega \doteq \frac{5.520}{r_0} \sqrt{\frac{T}{\sigma}}.$$

The solution is

$$z = f(0)J_0\left(\alpha_2 \frac{r}{r_0}\right) \cos \omega t \qquad \text{on} \quad [0, r_0].$$

The argument $\alpha_2(r/r_0) = \alpha_1$ when $r = (\alpha_1/\alpha_2)r_0$, and α_2 when $r = r_0$. The curve of a radial section of membrane on $[0, r_0]$ is a multiple of the J_0 curve on $[0, \alpha_2]$, compressed or stretched onto $[0, r_0]$ (see Fig. 3.9). Looking down on the membrane, there is a *nodal circle* of stationary points (where Chladni's sand collected).

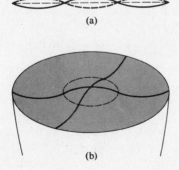

(a)

(b)   **Figure 3.9**

EXERCISES [3.14.1]
. . . .

1. Discuss the effects of changes in tension, radius, or density of a membrane. How do these effects compare with similar changes in a taut string?

2. What is the radius of the nodal circle in the second mode of vibration?

3. Describe in detail the vibrations in the third normal symmetric mode.

   a) Determine the frequency and nodal circles.

   b) Draw the curve $z(r, 0) = f(r)$, and make other sketches to illustrate the vibrations.

4. Discuss a few higher modes. Include drawings of nodal circles and sketches of vibrations to illustrate the discussion. The results may be checked with the pictures in Section 3.18.

5. a) Do the nodal circles of different symmetric modes coincide?

   b) Describe the simultaneous motion of the first two normal symmetric modes.

   c) Compute the frequencies of the first three normal symmetric modes as multiples of $(1/r_0) \sqrt{T/\sigma}$.

. . . .

For each integer $n > 0$, there is a zero α_n of J_0 and a solution

$$z = f(0)J_0 \left(\alpha_n \frac{r}{r_0} \right) \cos \omega t$$

describing the nth symmetric mode of vibration, with $n - 1$ nodal circles. The operators corresponding to both the wave equation and the ordinary equation for normal symmetric modes are linear. Therefore the sum of two solutions is a solution. The principle of superposition holds, and the membrane vibrates in different symmetric modes simultaneously. Just as there are Fourier series of sine and cosine functions, there are Fourier-Bessel series of Bessel functions.* But the Bessel zeros determine a sequence of frequencies which are not harmonic, that is, do not have a common period. Hence, the vibrations together do not form a repeating pattern, and produce a noise rather than a musical sound.

3.15 NORMAL MODES OF VIBRATION, THE BESSEL EQUATION AND J_1

Consider solutions of the form $z = f(r) \cos m\theta \cos \omega t$, which describe normal modes. In the absence of circular symmetry, another type of boundary condition must be imposed. In particular, $z = f(r, \theta, t)$ is a function, that is, is single valued, only if $z(r, \theta, t) = z(r, \theta + 2\pi, t)$. Thus m must be an integer. Any function $g = g(\theta)$ of period 2π satisfies this boundary condition, but any normal, separable solution has the form $z = f(r) \cos m\theta \cos \omega t$ (see Exercise 3 below).

EXERCISES [3.15.1]
. . . .

1. a) Determine and sketch nodal lines resulting from the $\cos m\theta$ term in the solution $z = f(r) \cos m\theta \cos \omega t$ for $m = 1, 2, 3$.

_____

*For lucid discussions of the Bessel solutions and their implications, see Bowman, Frank, *Introduction to Bessel Functions*, New York: Dover, 1958; Morse, P. M., *Vibration and Sound*, New York: McGraw-Hill, 1936.

b) Show that the same nodal patterns result from the term

$$A \cos m\theta + B \sin m\theta = C \cos(m\theta + \varphi)$$

in the solution

$$z = f(r) \, [A \cos m\theta + B \sin m\theta] \cos \omega t, \quad m = 1, 2, 3.$$

2. a) Show that for solutions of the form $z = f(r) \cos m\theta \cos \omega t$, the two-dimensional wave equation implies that $f(r)$ is a solution of the ordinary equation

$$f''(r) + \frac{1}{r} f'(r) + \left(\omega^2 \frac{\sigma}{T} - \frac{m^2}{r^2} \right) f(r) = 0, \qquad (3.15.1)$$

called a *Bessel equation*.

b) How is the Bessel equation related to Eq. (3.12.1), representing normal symmetric modes?

3. a) Continue the separation of variables started in Exercise 5 [3.12.1]. Separate f_1 and f_2, and by similar arguments obtain the two separate equations $f_2''(\theta) + m^2 f_2(\theta) = 0$ and

$$f_1''(r) + \frac{1}{r} f_1'(r) + \left(\omega^2 \frac{\sigma}{T} - \frac{m^2}{r^2} \right) f_1(r) = 0.$$

b) Is m restricted to integer values here, and if so why?

c) Compare the implications in the two approaches for simplifying the partial differential equation: the separation of variables, and the restriction of solutions in terms of motion and vibration patterns.

4. a) Let $m = 1$ in the Bessel equation in Exercise 3. Obtain a series solution and show that it may be written in the form

$$f(r) = A \sum_{k=0}^{\infty} (-1)^k \frac{1}{2^{2k+1} k! (k+1)!} \left(\sqrt{\frac{\sigma}{T}} \, \omega r \right)^{2k+1},$$

where

$$A = a_1 \, 2 \sqrt{\frac{T}{\sigma} \frac{1}{\omega}}.$$

b) Find the radius of convergence.

5. The function defined by

$$\sum_{k=0}^{\infty} (-1)^k \frac{1}{2^{2k+1} k! (k+1)!} x^{2k+1}$$

is called the *Bessel function of order 1* and is denoted as J_1.

a) Determine a numerical approximation of J_1 on $[0, 8]$ with three-place accuracy, using increments of 0.2.

b) Sketch J_1, the Bessel function of order 1.

c) Estimate and list the zeros of J_1 in $[0, 8]$.

6. Write the series for $J_0'(x)$ and $(xJ_1(x))'$, and determine relations between $J_0(x)$, $J_1(x)$, $J_0'(x)$, and $J_1'(x)$.

7. Discuss the similarities and differences between the two pairs of functions (J_0, J_1) and (sine, cosine). Include the spacing and number of zeros, as well as the periodicity and amplitudes.

. . . .

The Bessel equation (3.15.1) is a generalization of the ordinary equation (3.12.1) representing normal symmetric modes, and p has the same discontinuity. It is not surprising then that for $m = 1$, as for $m = 0$, the usual series method, without initial conditions, does not yield a general solution. In this case also, a general solution contains a logarithmic term which is not finite at $r = 0$. The boundary condition that z be finite for $r \leq r_0$ implies that the coefficient of this logarithmic solution in the general solution is zero. Thus, for $m = 1$, the separable solution for normal modes of vibration for the circular membrane is

$$z = 2\sqrt{\frac{T}{\sigma}}\frac{a_1}{\omega} J_1\left(\sqrt{\frac{\sigma}{T}}\omega r\right) \cos\theta \cos\omega t.$$

Both initial conditions are satisfied; $\cos\omega t$ indicates zero initial velocity and position $z(r, \theta, 0) = AJ_1(\sqrt{\sigma/T}\,\omega r)\cos\theta$. In this case, A must be determined by the initial displacement elsewhere than $r = 0$, since $J_1(0) = 0$.

The boundary condition $z(r_0, \theta, t) = 0$ acts as before; $J_1(\sqrt{\sigma/T}\,\omega r_0) = 0$ and $\sqrt{\sigma/T}\,\omega r_0$ must be a zero of J_1. The frequency ω can assume only the values $\sqrt{T/\sigma}(\alpha_i^1/r_0)$, where α_i^1 indicates the ith positive zero of J_1. The nodal curves arise from the zeros of both J_1 and the cosine function. As before, corresponding to each allowable value of ω there is a set of nodal circles. And in addition, for each mode, there is a nodal line $\theta = \pi/2$.

EXERCISES [3.15.2]
. . . .

1. a) Discuss the normal modes of vibration corresponding to $m = 1$ in the Bessel equation (3.15.1). Illustrate with sketches of nodal curves and positions of the membrane.
   b) Find a simple expression for the radius of the kth nodal circle in the nth mode.
   c) Compute frequencies $\omega_{1,n}$, $n = 1, 2, 3$, where n indicates the nth mode. Compare to the frequencies $\omega_{0,n}$ for the normal symmetric modes.

2. Determine series solutions for the Bessel equation for $m = 2, 3$, and the general case m. Discuss the difficulties involved.

. . . .

3.16 BESSEL FUNCTIONS, J_m

In the series solution of the Bessel equation of order m, the early coefficients a_k, $k < n$, are zero. It is convenient to assume a solution of the form

$$\sum_{k=0}^{\infty} a_k x^{k+c} = x^c \sum_{k=0}^{\infty} a_k x^k.$$

If the series is nonzero, it has a first term; hence a_0 may be assumed to be nonzero. Thus the coefficient of x^{c-2} determines the possible values of c (see Exercises 1 through 5 [3.16.1]). If c is real, this type of series enlarges, again, the class of possible solution functions, but introduces various problems. For example, if c is other than a positive integer, the series is not a power series. Is term-by-term differentiation valid? For the time being assume that $f(x) = \sum_{k=0}^{\infty} a_k x^{k+c}$ is well behaved and differentiable term-by-term, and determine the coefficients in Exercises [3.16.1]. These exercises will facilitate a discussion of the applicability of series methods in Section 3.19.

EXERCISES [3.16.1]
. . . .

1. Illustrative example: Consider the equation

$$4x\, y'' + 2(1 - x)\, y' - y = 0.$$

   a) Write the equation in the form $y'' + p(x)\, y' + q(x)\, y = 0$.
   b) Obtain solutions of the form $y = \sum_{k=0}^{\infty} a_k x^k$. Does this method yield a general solution?
   c) Assume termwise differentiable solutions of the form $\sum_{k=0}^{\infty} a_k x^{k+c}$. Substitute the series in the differential equation and show that the coefficient of x^{c-2}, the lowest power of x, is $a_0 c(4c - 2)$. The condition $c(4c - 2) = 0$ is called the *indicial equation*.
   d) Use the roots c_i of the indicial equation to determine specific series solutions, one for each c_i. Find the radii of convergence. Are these solutions independent? (Remember, to show independence on I, you need only find a single value at which the Wronskian is zero.)
   e) Write a general solution, if possible.

2. a) Determine the indicial equation for

$$y'' + \frac{(1 + x)}{2x}\, y' - \frac{1}{x}\, y = 0.$$

   b) Determine the recurrence relation for each c_i.
   c) Compute the first five coefficients for each series.
   d) Write a general solution, if possible.

3. Determine the indicial equation for the Bessel equation of order 2, 3, 4, and m.

4. The Bessel equation of order m, in standard form, is

$$y'' + \frac{1}{x} y' + \left(1 - \frac{m^2}{x^2}\right) y = 0.$$

[The constant ω^2 (σ/T) is replaced by 1.]
a) Obtain recurrence relations for $m = 2, 3$ for each c_i.
b) Determine solutions for $m = 2, 3$.
c) Write general solutions for $m = 2, 3$, if possible.
d) Obtain a solution for each c_i for the equation of order m.
e) Show that a solution of the mth-order equation may be written as

$$y(x) = 2^m m! \sum_{k=0}^{\infty} (-1)^k \frac{x^{m+2k}}{2^{2k} k! \, (m + k)!}.$$

The multiplicative factor is arbitrary in a solution of a linear homogeneous equation. For convenience in other relations, the *Bessel function of order m* (m a nonnegative integer) is defined as the particular solution

$$J_m(x) = \sum_{k=0}^{\infty} (-1)^k \frac{x^{m+2k}}{2^{2k} k! \, (m + k)!}.$$

By definition, $J_{-m}(x) = (-1)^m J_m(x)$. Does this agree with the solution for $c = -m$ in Exercise 4(d), except for a multiplicative constant? Numerical values and zeros of J_2, J_3, J_4 may be calculated as before. It is enlightening to look at the curves and the tables of values and approximate zeros of J_m for various values of m.*

EXERCISES [3.16.2]

1. Write power series solutions for the Bessel equation

$$f''(r) + \frac{1}{r} f'(r) + \left(\omega^2 \frac{\sigma}{T} - \frac{m^2}{r^2}\right) f(r) = 0, \qquad \text{for} \quad m = 2, 3, 4.$$

2. Write solutions of the form $z = f(r) \cos m\theta \cos \omega t$, $m = 2, 3, 4$, for the wave equation of the circular membrane.

3. Discuss and sketch nodal curves for solutions corresponding to $m = 2, 3, 4$. Discuss the nodal curves of $z = f(r) \cos m\theta \cos \omega t$.

---

*Jahnke, Eugene; Emde, Fritz; Losch, Friedrich, *Tables of Higher Functions*, 6th ed., 132–134, 158–163, 192–195, New York: McGraw-Hill, 1960.

4. Compute and compare the frequencies,

$\omega_{0,1}, \omega_{0,2}, \omega_{0,3}$, for the first three normal symmetric modes,

$\omega_{1,1}, \omega_{1,2}, \omega_{1,3}$, for normal modes with one nodal diameter,

$\omega_{2,1}, \omega_{2,2}, \omega_{2,3}$, for normal modes with two nodal diameters.

. . . .

3.17 CHARACTERISTIC VALUES (OR EIGENVALUES) AND CHARACTERISTIC FUNCTIONS (OR EIGENFUNCTIONS)

Each function

$$z = AJ_m\left(\alpha_n^m \frac{r}{r_0}\right) \cos m\theta \cos \omega t; \qquad m = 0, 1, 2, \ldots, n = 1, 2, \ldots,$$

satisfies the two-dimensional wave equation and the boundary conditions

1) $z(r_0, \theta, t) = 0$;

2) z is bounded for $r \leq r_0$; and

3) $z(r, \theta, t) = z(r, \theta + 2\pi, t)$.

Each of these solutions represents a normal mode of vibration with frequency

$$\omega_{m,n} = \sqrt{\frac{T}{\sigma}} \frac{\alpha_n^m}{r_0}$$

where α_n^m is the nth positive zero of J_m. The discrete frequencies $\omega_{m,n}$ are called characteristic values, or eigenvalues (and are defined to within an arbitrary multiplicative constant). The functions

$$\psi_{m,n}(r, \theta) = J_m\left(\alpha_n^m \frac{r}{r_0}\right) \cos m\theta$$

are called the characteristic functions or eigenfunctions of the boundary value problem. The term $\psi_{m,n}$ represents a vibration pattern with m nodal diameters and $n - 1$ nodal circles. This is but one example of this particular classification of solutions of a boundary-value problem. In general, the eigenfunctions characterize a restricted set of solutions representing normal modes of vibration.

The remarks at the end of Section 3.14 concerning superposition, simultaneous vibrations, Fourier-Bessel series, and the interference of nonharmonic vibrations, apply equally well to this set of characteristic functions. The frequencies are not harmonic.

3.18 VIBRATING MEMBRANES: EXPERIMENTAL RESULTS

Do membranes actually vibrate as predicted by these solutions? The first experimental studies were made by Chladni (1756–1827) of Saxony. Chladni trained for law, but turned to music and is now considered the first student of modern acoustics. The following excerpt from the autobiographical introduction to his *Die Akustik*, published in 1802, appears in translation in Tyndall's book on sound.*

As an admirer of music, the elements of which I had begun to learn rather late, that is, in my nineteenth year, I noticed that the science of acoustics was more neglected than most other portions of physics. This excited in me the desire to make good the defect, and by new discovery to render some service to this part of science. In 1785 I had observed that a plate of glass or metal gave different sounds when it was struck at different places, but I could nowhere find any information regarding the corresponding modes of vibration. At this time there appeared in the journals some notices of an instrument made in Italy by the Abbe Mazzochi, consisting of bells, to which one or two violin-bows were applied. This suggested to me the idea of employing a violin-bow to examine the vibrations of different sonorous bodies. When I applied the bow to a round plate of glass fixed at its middle it gave different sounds, which, compared with each other, were (as regards the number of their vibrations) equal to the squares of 2, 3, 4, 5, etc.; but the nature of the motions to which these sounds corresponded, and the means of producing each of them at will, were yet unknown to me. The experiments on the electric figures formed on a plate of resin, discovered and published by Lichtenberg, in the memoirs of the Royal Society of Gottingen, made me presume that the different vibratory motions of a sonorous plate might also present different appearances, if a little sand or some other similar substance were spread over the surface. On employing this means, the first figure that presented itself to my eyes upon the circular plate already mentioned resembled a star with ten or twelve rays, and the very acute sound, in the series alluded to, was that which agreed with the square of the number of diametrical lines. (See Fig. 3.10.)

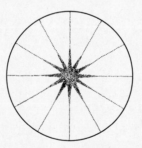

Figure 3.10

Chladni gave lecture demonstrations in the courts of Europe and so delighted Napoleon that the emperor financed a French translation of *Die Akustik*. This work contains nodal patterns for plates of various shapes, each excited by bowing and damping at different points. There are 43 nodal diagrams for the circular plate alone. Chladni also developed a tonemeter to measure the frequencies of the

---

*Tyndall, John, *Sound*, 3rd ed., New York: D. Appleton & Co., 1885.

vibrations. Excellent accounts of Chladni's work are given by Tyndall and by Waller.*

In addition to acoustics, Chladni initiated the modern study of meteorites.† Meteorites were reported, along with other miracles in ancient times (see Joshua 10:11 in the Old Testament of the *Bible*), and stones had been preserved. But men of science in the seventeenth and eighteenth centuries rejected miracles, including stones from heaven. When, in 1794, Chladni published his conclusion that iron masses did indeed come from outside the earth, he was ridiculed and accused of evil. A well-observed fall in France in 1803 lent support to Chladni's studies.

In nineteenth-century treatises on acoustics, the Chladni figures were cited in support of the theoretical solutions for vibrating membranes. Occasional mention was made of the discrepancies between predicted and measured frequencies. Actually, Euler's differential equations and solutions for the circular membrane and Chladni's nodal patterns for plates concern two distinct though related problems. Plates have a stiffness of their own and are subject to elastic (restoring) forces similar to those described by Hooke's law. Membranes are flexible; the elastic forces are negligible compared to the tension forces. A comparison of the theoretical and physical aspects of these two objects clarifies each individual situation. (A similar relation exists between vibrating strings and rods.)

Both membranes and plates have been long and widely used in acoustics as *transducers*, devices that allow energy to change from one form to another. For example, most musical instruments are acoustical-mechanical transducers, that is, they change mechanical energy into sound energy (sound waves). In the drum, the transfer is made via the vibrating membrane; in cymbals, the transfer is made via vibrating plates. Bells and violin surfaces are just bent plates. Microphones, loudspeakers, and telephones are acoustical-electrical transducers. In a telephone, the essential part is a vibrating plate with circular boundary; in a microphone, it is a vibrating membrane. In transforming sound to electrical energy, the plates or membranes are excited by impressed forces. The human eardrum, an oval-shaped membrane about 0.07 mm thick, is stretched over a bone frame; it is the first of a sequence of devices that convert impressed sound waves into electrical impulses in the brain.

The early experiments involved plates, not membranes. It is easier to fabricate plates with uniform thickness and density; and it is easier to support plates and to design controlled and repeatable vibration experiments for them. Chladni's patterns and frequency measurements of the eighteenth century are still valid.

The mathematical model (3.11.1) for the membrane is simpler. The elastic forces in plates, even with simplifying assumptions equivalent to those for

---

*Waller, Mary D., *Chladni Plates*, London: Staples Press, 1960.

†Heide, Fritz, *Meteorites*, *Phoenix Science Series*, Translation of 2nd ed., Chicago: U. of Chicago Press, 1957.

membranes, lead to fourth-order differential equations. It is not surprising that the early mathematical models were developed for membranes. Present theory predicts that the nodal vibration patterns are the same for plates and membranes; this has been observed for some time. These common symmetries have also received algebraic treatment.* The frequencies, however, depend on the stiffness factor of the material.

The problem of measuring amplitudes is difficult in both situations. Remember that one of the simplifying assumptions, both for strings and membranes, is that the amplitudes are small, so small that sines and tangents of the relevant angles are equal. (For the first normal symmetric mode, this means roughly that the ratio amplitude/radius = 0.12.) Larger amplitudes result in a different kind of motion; the tension is no longer constant, the corresponding equation is nonlinear. But amplitudes small enough to justify the use of the simplifying assumptions and a linear equation are extremely difficult to measure. Only in 1968, was a suitable technique for such measurements developed. Indeed, all the post-Chladni developments in physical techniques for measuring these vibrations are quite recent.

In 1932, an ice-cream vendor outside a London medical school asked Dr. Mary Waller why his bicycle bell rang when he touched it with dry ice. During the next decade, Dr. Waller† developed the carbon-dioxide method of producing Chladni figures. This method increased the number of ways in which the plates can be excited, and extended the study of vibration patterns. Robinson and Stephens‡ in 1934, and Bergmann§ in 1956, produced and photographed vibrating soap films. Ordinarily, a soap film is not of uniform thickness: uniform thickness is obtained by rotating a horizontal film so that the excess liquid moves to the edge of the membrane. The vibrations were produced by loudspeakers, that is, the soap films vibrated under impressed forces. In the photographs of normal symmetric modes of vibration (Fig. 3.11), the antinodes of vibration appear as bright zones on a darker background. A vibrating circular soap membrane in stroboscopic illumination, as well as the nodal lines of one mode of vibration of a square membrane, is shown in Fig. 3.12. Soap films 1.5 cm in radius exhibited amplitudes of about 1.5 mm at the lowest frequencies.

Holographic interferometry was developed at the University of Michigan in 1965. It was applied to vibrating surfaces by Monahan and Bromley‖ in 1968.

---

*Melvin and Edwards, "Group Theory of Symmetric Molecules, Membranes, and Plates," *J. Acoust. Soc. Am.* **28**, 201 (1956).

†Waller, Mary D., *Chladni Figures*, London; G. Bell & Sons, 1961.

‡Robinson, N. W. and Stephens, R. W. B., "On the Behaviour of Liquid Films in a Vibrating Air Column," *Phil. Mag.*, **17**, 27–33 (1934).

§Bergmann, L., "Experiments with Vibrating Soap Membranes," *J. Acoust. Soc. Am.*, **28**, 1043 (1956).

‖Monahan, M. A. and Bromley, K., "Vibration Analysis by Holographic Interferometry," *J. Acoust. Soc. Am.*, **44**, 5, 1225–31 (1968).

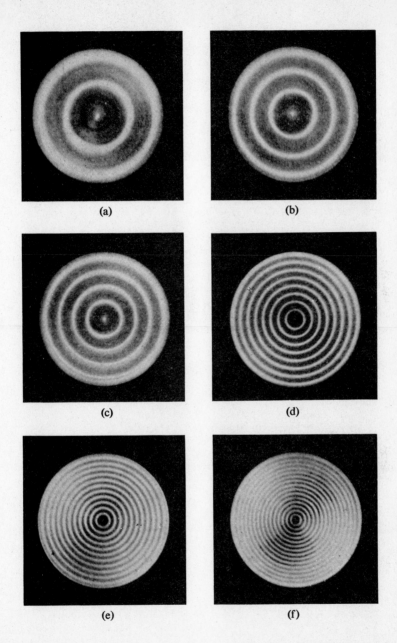

(a) (b) (c) (d) (e) (f)

Fig. 3.11 Vibrations of a circular soap membrane. (a)–(c) 3rd to 5th normal symmetric modes (or 2nd to 4th harmonic vibrations) (d)–(f) 9th, 12th, and 16th normal symmetric modes. In these photographs, the white regions indicate antinodes or regions of maximum amplitude. Dark regions correspond to annuli about the nodal circles. (Nodalrings at the edge of the membrane are difficult to see because of the background.) (L. Bergmann, *J. Acoust. Soc. Amer.* **28,** 6, November 1956, page 1043f)

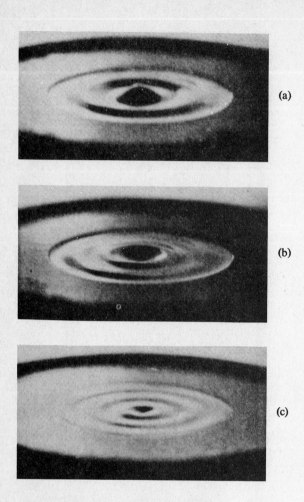

(a)

(b)

(c)

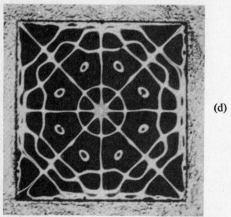

(d)

Fig. 3.12 (a)–(c) Vibrating circular soap membranes in stroboscopic illumination. (a) and (b) 4th normal symmetric mode; (c) 5th normal symmetric mode. (d) A photograph of one mode of vibration of a square soap membrane. Note the extraordinary symmetry of this vibration pattern. (L. Bergmann, *J. Acoust. Soc. Amer.* **28**, 6, November 1956, page 1043f)

The technique will not be described in detail here. Only a simplified and intuitive description, sufficient to interpret the resulting photographs, will be given. Briefly, a pattern of light formed by interfering beams of coherent light (in phase) reflected from a vibrating surface is recorded. In sinusoidal motion, a point is within 25% of its maximum displacement 67% of the time. The recorded pattern of light indicates the difference in distance traveled by light reflected from the far and near positions of the vibrating surface. If the difference, for a particular point on the surface is an even multiple of $\lambda/2$ (where λ is the wavelength of the light used), then the light is reinforced and the optical density recorded is high. If the difference is an odd multiple of $\lambda/2$, then the optical density is low.

The pattern in Fig. 3.13(a) is a contour map of the amplitudes of vibration. The light source is a helium-neon gas laser with wavelength 0.6328 μ. (A micron, μ, is 10^{-6} m.) The vibrating surface is a sonar transducer. The points on the boundary curve are stationary; the light density at points corresponding to the edge of the membrane is high. The first dark fringe indicates a phase difference of $\lambda/2$; the two light beams negate each other. The difference in the two extreme positions of points on the membrane, corresponding to points on this first dark fringe of the photograph, is $\lambda/4 \doteq 0.1582\ \mu$. The amplitude of vibration for points on this fringe (taking errors into account) is about $0.08 \pm 0.010\ \mu$. The next light fringe indicates an amplitude of about $0.16 \pm 0.010\ \mu$. Thus the peak amplitude is about $0.32 \pm 0.010\ \mu$ in the upper lobe and $0.24 \pm 0.10\ \mu$ in the lower lobe.

This technique of holographic interferometry, then, can measure amplitudes of the order of 10^{-7}m, and indicate in a single hologram recording the amplitude at every point on a vibrating surface.

None of the photographs in Figs. 3.13 and 3.14 represent normal modes of vibration, although they are all close to the normal mode with one diametral nodal line. The sequence of photographs in Fig. 3.13 show that as the driving voltage, or impressed force, is increased, the amplitudes increase as expected. In Fig. 3.14, where the driving force is constant and the frequencies increase, the greatest amplitude and symmetry occur in C. This suggests that the characteristic frequency $\omega_{1,1}$, at which the membrane would resonate in a normal mode, is just below 593.7 Hz (Hertz is 1 cycle/sec.) In both figures the energy shifts from lobe to lobe as the driving force, or frequency, increases. The amplitudes measured by this method support the Bessel function model.

The overtones of a membrane are nonharmonic, the sound should be non-musical. What about the exceptions? The tabla, a pair of drums long used in northern India, and the south Indian mridanga produce musical sounds. The membranes are loaded with a black paste (sometimes made of manganese dust, boiled rice, and tamarind juice; other times made from iron oxide, charcoal, starch, and gum) which hardens and yet remains flexible.* The load appears as a black disk

*Cohen, H. and Handelman, G., "On the Vibration of a Circular Membrane with Added Mass," *J. Acoust. Soc. Am.*, **29**, 2, 1957.

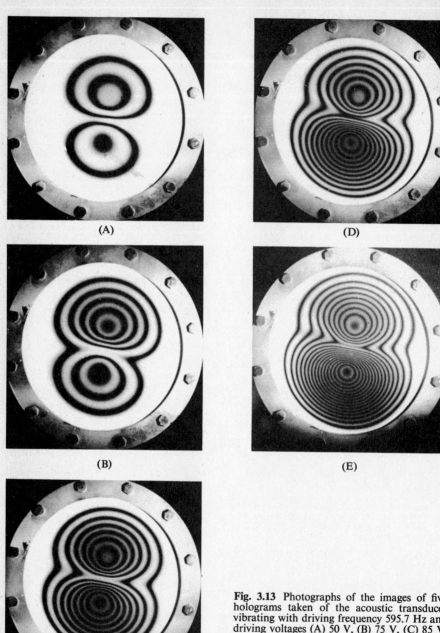

(A)

(D)

(B)

(E)

(C)

Fig. 3.13 Photographs of the images of five holograms taken of the acoustic transducer vibrating with driving frequency 595.7 Hz and driving voltages (A) 50 V, (B) 75 V, (C) 85 V, (D) 100 V, and (E) 125 V.   The patterns (or contour lines) suggest one nodal diameter and no nodal circles.   (M. A. Monahan and K. Bromley, *J. Acoust. Soc. Amer.* **44**, 5, November 1968, page 1225–31)

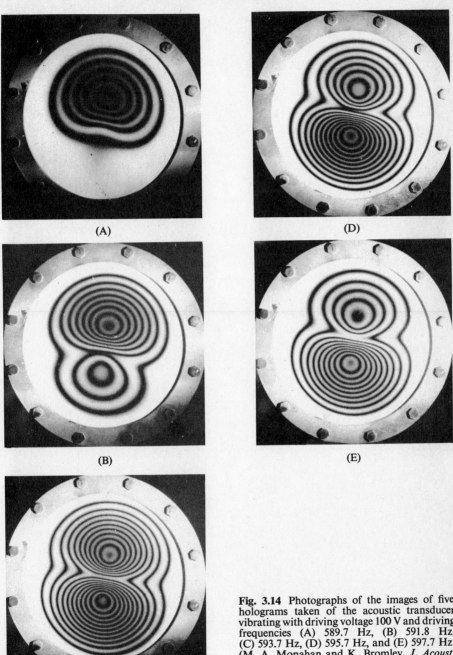

Fig. 3.14 Photographs of the images of five holograms taken of the acoustic transducer vibrating with driving voltage 100 V and driving frequencies (A) 589.7 Hz, (B) 591.8 Hz, (C) 593.7 Hz, (D) 595.7 Hz, and (E) 597.7 Hz. (M. A. Monahan and K. Bromley, *J. Acoust. Soc. Amer.* **44,** 5, November 1968, page 1225–31)

(a)

(b)

Figure 3.15

on the membrane (see Fig. 3.15a). The symmetric load on the right-hand drum of the tabla is precisely that which produces harmonic overtones (see Exercise 1 [3.18.1]).* The left-hand drum is asymmetrically loaded; a musical tone is produced when the heel of the hand is pressed into the vibrating membrane to create a symmetric load (see Fig. 3.15b). A recent analysis by Ramakrishna shows that membranes may be loaded so that the overtones are altered while the fundamental frequency remains the same.† His analysis is based on numerical approximations to solutions of a nonseparable equation, a technique made possible by high-speed computers and recent developments in numerical methods.

EXERCISES [3.18.1]

1. Suppose a concentric load on a membrane varies inversely as r^2, that is, it is heaviest at the center, and for $r \geq \varepsilon$, $\sigma = \sigma_0 r^2$.

   a) Show that the wave equation is

   $$\frac{\sigma_0}{T} z_{tt} = r^2 z_{rr} + r z_r + z_{\theta\theta}.$$

   b) Write the ordinary equation corresponding to the normal symmetric modes.
   c) Show that $f(r) = A \cos(\omega \sqrt{\sigma_0/T} \ln r)$ is a solution of the ordinary equation and the usual initial conditions.
   d) Determine the characteristic values for the boundary condition $z(r_0, \theta, t) = 0$.
   e) Show that the overtones are harmonic.

2. Consider a square membrane with the usual initial conditions.

   a) Assume separation of variables and determine the ordinary equations for the normal modes.
   b) Determine nodal curves and frequencies $\omega_{m,n}$ for $m = 1, 2, 3$, $n = 1, 2, 3$.

3.19 SOME REMARKS ON SERIES SOLUTION METHODS

The series solutions J_m, $m = 0, 1, 2, \ldots$, of the Bessel equation have been useful in determining characteristic functions for the circular membrane. How widely applicable are series solution methods? Theorem 2.1.1 for second-order linear equations asserts the existence of a solution, satisfying initial conditions, if the

---

*Ghosh, R. N., Note on Indian Drums, *Phys. Rev.* **20**, 526 (1922).

†Ramakrishna, B. S., "Modes of Vibration of the Indian Drum Dugga or Lefthand Thabala," *J. Acoust. Soc. Am.*, **29**, 2 (1957). Ramakrishna, B. S. and Sondhi, M. M., "Vibrations of Indian Musical Drums Regarded as Composite Membranes," *J. Acoust. Soc. Am.*, **26**, 523–529 (1954).

coefficient functions p and q are continuous. Yet the Bessel equation has coefficients with discontinuities at 0, and still has series solutions defined at 0. Is continuity necessary? If not, how bad can the discontinuities be?

The process of obtaining series solutions for linear equations with simple polynomial coefficients such as 0, 1, or x (Section 3.10) suggests that series methods are feasible for equations with polynomial coefficients; coefficients x^k, k an integer, are easily absorbed in the trial series by termwise multiplication. Polynomials of higher degree merely increase the number of operations. What other types of coefficients can be absorbed in, or combined with, the trial series? The natural extension of a polynomial is a power series. Multiplication is well defined for power series on a common interval of absolute convergence. Although the procedure is complicated, each a_k of the solution series is determined by the preceding $a_j, j < k$, and the operations may be programmed for a computer. Thus a numerical approximation may be obtained. Indeed, there is an existence theorem (which will not be proved here) that asserts a power series solution $\sum_{k=0}^{\infty} a_k x^k$ for a linear differential equation whose coefficients have power series representations at $x = 0$, and initial conditions $y(0) = y_0$ and $y'(0) = y'_0$, and the power series solution may be specified by the method of undetermined coefficients. Since a power series at 0 is differentiable on $(-\rho, \rho)$, the class of equations with power series coefficients is a subclass of those with continuous coefficients. In this case, the stronger hypothesis nets a method of solution.

Certain types of discontinuities do not preclude series solutions. In particular, equations of the form

$$y'' + \frac{p(x)}{x} y' + \frac{q(x)}{x^2} y = 0, \tag{3.19.1}$$

where $p(x)$ and $q(x)$ have power series representations at 0, have solutions of the form $|x|^c \sum_{k=0}^{\infty} a_k x^k$; the possible values of c are determined by the indicial equation (see Exercise 1(c) [3.16.1]). Frobenius (1849–1917) introduced the series $\sum_{k=0}^{\infty} a_k x^{k+c}$ as a solution on a deleted interval $I - \{0\}$. The differential equation (3.19.1) is said to have a *regular singular point at* 0. In this case, the hypothesis is weaker, the coefficients need not be continuous; but a method of solution, the Frobenius method, is applicable. However, although solutions may be determined, they may not satisfy arbitrary initial conditions, that is, the method does not always yield a general solution. If the roots c_i of the indicial equation differ by a noninteger, the corresponding solutions are independent. Otherwise the solutions may be dependent; but even in this case, other methods have been developed for finding independent solutions. One of these methods involves the introduction of a logarithmic function. For example, the general solution of the Bessel equation of order 0 may be written as

$$AJ_0(x) + B\left[J_0(x) \ln x + \sum_{k=0}^{\infty} b_k x^k\right] = AJ_0(x) + BY_0(x).$$

Since the drumhead attains only finite heights for $0 \leq r \leq r_0$, the boundary conditions are met only if $B = 0$. An annular membrane (bounded by two concentric circles) might well have a solution with $B \neq 0$.

The techniques and theory concerning solutions of linear differential equations with regular singular points may be found in:

Greenspan, Donald, *Theory and Solution of Ordinary Differential Equations*, New York: Macmillan, 1960.

Coddington, Earl A., *An Introduction to Ordinary Differential Equations*, Englewood Cliffs, N.J.: Prentice Hall, Inc., 1961.

Rainville, Earl D., *Intermediate Differential Equations*, London: Wiley & Sons, 1943.

Rainville, Earl D., *Special Functions*, New York: Macmillan, 1960.

Many of the differential equations of physics and applied mathematics have regular singular points at real values x_0 (not necessarily zero). Indeed, the Bessel equation itself is a mathematical model for many different physical situations. And different pairs of independent solutions, each including J_m, seem particularly suitable for certain physical situations. There is no end to the possible varieties of these *Bessel functions of the second kind*. Rephrased, two independent solutions provide a basis; all other solutions may be expressed as a linear combination of them. There are many different bases. It is possible that a familiar function may be unrecognizable when expressed in terms of a particular basis (see Exercise 4 below). This situation is essentially that discussed in Chapter 1; general solutions may be useful, but if initial conditions are not imposed, the variety of solutions may be confusing.

Do series solution methods apply to nonlinear equations? If $y = \sum_{k=0}^{\infty} a_k x^k$, how are $(y')^{1/2}$ or $(y'')^3$ evaluated? The first is beyond our range of techniques. The second is possible, using multiplication of series, but leads to recurrence relations involving roots too complicated to be practical. On the other hand, series methods are applicable to higher-order linear equations (see Exercise 2 below).

In conclusion then, series methods are applicable to linear equations whose coefficients have power series expansions, or whose discontinuities are limited to regular singular points. Their use in the case of irregular singular points is not neatly characterized.

★EXERCISES [3.19.1]

. . . .

1. Write a general linear differential equation of order 2 with a regular single point at x_0. What type of solution is probable?

2. a) Write a general linear differential equation of order 3 with a regular single point at 0. What type of solution is probable?

b) Discuss relations among the indicial equation, recurrence relations, independence, the Wronskian, and general solutions.

3. a) Show that the indicial equation of the equation

$$y'' + \frac{p_0}{x} y' + \frac{q_0}{x^2} y = 0,$$

where the coefficients $p(x)$ and $q(x)$ are constants p_0 and q_0, is

$$c^2 + (p_0 - 1)c + q_0 = 0.$$

b) Show that the same indicial equation holds for the equation

$$y'' + \frac{1}{x} \sum_{k=0}^{\infty} p_k x^k y' + \frac{1}{x^2} \sum_{k=0}^{\infty} q_k x^k y = 0,$$

where the coefficient functions p and q have power series expansions

$$p(x) = \sum_{k=0}^{\infty} p_k x^k \quad \text{and} \quad q(x) = \sum_{k=0}^{\infty} q_k x^k.$$

4. a) Determine independent series solutions of $y'' - 3y' + 2y = 0$.

b) Do you recognize the series? Compare the general solutions obtained by algebraic and series methods.

c) Impose initial conditions and apply algebraic and series methods.

d) Write the general solution, obtained by algebraic methods, in series form. Recombine these series to form other pairs of independent solutions, which might not be recognized as series representations of familiar functions.

. . . .

3.20 SOME REMARKS ON THE HISTORY AND TERMINOLOGY OF BESSEL FUNCTIONS

The astronomer Bessel (1784–1846), at age 15, was placed in a counting house in Bremen. He wanted to go to sea and visit foreign lands. So in his spare time he studied navigation, and soon became involved in mathematics and astronomy. At 20, the publication of his calculation of the orbit of Halley's comet led to a post at an observatory. After his masterly investigation of the comet of 1807, he was summoned by the King of Prussia to direct the new observatory at Königsberg, where he remained until his death. It is interesting to note that he found it necessary, in 1826, to correct the length of the seconds pendulum then in use.

In 1824 Bessel's paper concerning planetary perturbations contained the first numerical tables of J_0 and J_1, on the interval $[0, 3.2]$ with increments of 0.01, and 10 decimal accuracy. The functions had already been used by the Bernoullis,

Euler, Lagrange, Fourier, and Poisson. Euler's 1764 paper on the vibrating membrane presented the two-dimensional wave equation in polar coordinates, the Bessel equation of order m (3.15.1) for the normal modes of vibration, and the solutions AJ_m, m an integer. In 1769, Euler introduced the function Y_0 of the second kind; one evaluation of Y_0 involves Euler's constant γ (Exercise 6 [3.2.2]).

Bessel himself introduced an integral form for the functions J_m and a notation that greatly facilitated their study. And Jacobi, in response to this ingenuity, referred to the J_m as Bessel functions, in an 1836 publication.

The $J_m(x)$, m an integer, are often called Bessel coefficients because of the relation

$$\exp\left[\tfrac{1}{2}x\left(t - \frac{1}{t}\right)\right] = \sum_{k=-\infty}^{\infty} J_k(x)t^k.$$

In this series representation, established about a decade after Bessel's death, the coefficients J_k involve the factorial function which is defined only for nonnegative integers. But Euler's gamma function is a generalization of $n!$, that is, $\Gamma(z)$ is defined for reals and even complex numbers; for $z = n$, a positive integer, $\Gamma(n + 1) = n!$. Just so, the Bessel coefficients were generalized, first to real orders by Lommel in 1868, and in the next year to complex orders by Hankel. Now, Bessel functions indicate functions of complex order n on a complex domain.

As Watson states in his *Theory of Bessel Functions*, "a large part of the theory of Bessel functions has been constructed expressly for the purpose of facilitating numerical computations."* These functions have played a significant role not only in astronomy but in such diverse fields as the theory of sound, hydronamics, diffraction of light, tidal waves, the propagation of electromagnetic waves along wires, and the flow of heat or electricity in cylinders. Numerical values were in demand long before the advent of high-speed computers.

The same problems, applied to different shapes, give rise to boundary conditions expressed in different coordinates, and yield other well-known special functions as solutions. For example, coordinates for circular boundaries lead to Bessel functions; parabolic boundaries yield Hermite functions. Both types of problems are special cases of the Dirichlet problem, that is, the problem of finding a solution of a second-order partial differential equation in two variables if the function values are specified on a closed curve. These partial equations with boundary conditions often lead to second-order linear equations. The basic theorem for second-order linear equations, existence theorem 2.1.1, has been stated and used, but not proved. In Chapter 4, this proof will be presented as an extension of the existence and uniqueness theorem for the first-order equation $y' = f(x, y)$.

---

*Watson, G. N., *A Treatise on the Theory of Bessel Functions*, 2nd ed., New York: Macmillan, 1944.

CHAPTER FOUR

EXISTENCE THEOREMS

4.1 INTRODUCTION

The existence theorems in Chapter 1 provide specific methods for writing solutions. The solutions, uniquely defined by integral equations, may be expressed in terms of familiar antiderivative functions, or approximated by numerical tables.

The differential equation $y' = f(x, y)$ includes as special cases all the equations considered in Chapter 1. It also includes equations which fit none of the previous classifications. Is it possible to formulate an existence and uniqueness theorem for such a general equation? If so, a theorem with a weaker conclusion and/or a stronger hypothesis than that for the separable equation must be expected. Is there a method for obtaining a solution, or a simple form for writing the solution, in all instances? Actually there are several theorems concerning this equation. The hypotheses vary in strength and the mathematical arguments vary in complexity. The theorem and proof presented here fall within the framework of the calculus. Only a few concepts not usually included in first-year calculus are needed: closed sets, uniform convergence, and some implications of continuity.

As before, the set of solutions is restricted by considering the differential equation together with an initial condition. The theorem guarantees a unique local solution containing (x_0, y_0). The hypothesis is stronger; continuity is not sufficient. Lipschitz (1832–1903) introduced the additional condition on f, used here. The conclusion is weaker; no method for writing the solution in usable form is provided. The proof is constructive, that is, the existence is again established by producing a solution. But this solution, given by Émile Picard (1856–1941) in the 1890's, is defined as the limit of a sequence of integral functions, and often cannot be expressed in any other form.

Lipschitz conditions and Picard approximation functions are met in other situations. These same methods may be applied to systems of differential equations and higher-order equations. In particular, existence theorem 2.1.1 for second-order linear equations may be proved by these methods. And the Picard approximations generate numerical methods known as finite difference schemes.

The final section of this chapter is a résumé and comparison of numerical methods.

4.2 AN OUTLINE OF THE PROOF

The proof is presented leisurely. The theorem, which might be forbidding now, is not stated until after the new concepts have been introduced and developed in the exercises along with the essential arguments. To clarify this procedure a brief outline of the proof is given.

1. The solution function is characterized by an integral equation. Specifically, a continuous function $s: x \rightarrow y$ on I is a solution of $y' = f(x, y)$ and $y(x_0) = y_0$, if, and only if, s satisfies the integral equation

$$y(x) = y_0 + \int_{x_0}^{x} f(t,\ y(t))\ dt \quad \text{(Sections 4.3 and 4.4)}.$$

2. This integral equation is circular; it does not specify a solution. But it does provide the structure for the Picard integral functions which approximate a solution. For example, suppose y_0 replaces $y(t)$ in the integral equation, and

$$y_0 + \int_{x_0}^{x} f(t,\ y_0)\ dt$$

defines a number, call it $s_1(x)$, for each $x \in I$; and $s_1(t)$, replacing y_0 in the integrand, defines values $s_2(x)$. In this way a sequence of functions $\langle s_n \rangle$ may be defined, where

$$s_n(x) = y_0 + \int_{x_0}^{x} f(t,\ s_{n-1}(t))\ dt.$$

These Picard integral functions have a special property: each successive function is a better approximation to a solution of the integral equation than its predecessor (Sections 4.5 through 4.7).

3. The Picard functions have a common domain of definition (Sections 4.8 and 4.9).

4. A Lipschitz condition on f ensures the convergence of the sequence of Picard functions. The convergence is established by the comparison test (Section 4.10).

5. The limit function is continuous and satisfies the integral equation. Thus it is a solution of the differential equation. Moreover, the same Lipschitz condition implies that the solution is unique (Sections 4.11 and 4.12).

6. The theorem is formally stated in Section 4.13.

Students from different disciplines read differential equations, and their tolerance for existence theorems varies. Much of the development here is in the exercises, to be done by the student. A science student with small tolerance for theory might get the main ideas by developing the Picard sequence for a specific equation and working through the convergence arguments. The equivalence of the integral equation, and the common domain for the Picard integral functions, might be assumed. Specifically, this would mean working through Exercises [4.6.1], [4.10.1], and [4.10.2] carefully.

Mathematics students should note the impact of continuous functions throughout the development. These concepts of continuity, generalized, are the basis of point-set topology.

4.3 THE EQUATION $y' = f(x, y)$

Let f be defined on a region $P = \{(x, y)\}$ in the plane. Do the equations

$$y'(x) = f(x, y(x)) \quad \text{and} \quad y(x_0) = y_0$$

determine a unique solution s on some interval $I \supset \{x_0\}$? If so, $y'(x_0) = f(x_0, y_0)$ must be well defined, that is, (x_0, y_0) must be in P. Indeed, if the differential equation is to be satisfied, the value $f(x, y(x))$ must be defined for each point $(x, y(x))$, $x \in I$. That is, the graph $\{(x, y(x)): x \in I\}$ of any solution function must be in P. Therefore the set of solution functions is contained in the set

$$\mathscr{F} = \{w \text{ on } I_w : \{(x, w(x)): x \in I_w\} \subset P\}$$

Rephrased, $\mathscr{F}$ is the set of functions whose graphs are in P.

In Chapter 1, the solution of each special case of this general differential equation was defined by an integral equation. That is, each system of a differential equation and initial equation was replaced by an equivalent integral equation (that is, their solution sets are identical). For the equation $y'(x) = f(x, y(x))$, with initial condition, an equivalent integral equation again plays the crucial role. This time it does not define the solution function uniquely, but provides the basic structure for the definition of the Picard integral functions. This integral equation has the same form as the solution in Theorem 1.3.1. Indeed, the equivalence may be established as a corollary of Theorem 1.3.1. As a first step, $y'(x) = f(x, y(x))$ will be interpreted as an equation of the form $y'(x) = \bar{f}(x)$.

EXERCISES [4.3.1]

. . . .

1. Write specific equations of the form $y' = f(x, y)$ which are
   a) separable;
   b) nonseparable, but linear;
   c) neither separable nor linear, but homogeneous (Section 1.18);
   d) nonseparable, nonlinear, and nonhomogeneous.

2. Write a first-order differential equation which is not of the form $y' = f(x, y)$. Note that the most general first-order equation is of the form $F(x, y(x), y'(x)) = 0$.

3. Describe or sketch the sets of points $\{(x, y, z)\}$ that satisfy

a) the equation $z = f(x, y) = 2 - x - y$, $|x| \le 2$, $|y| \le 2$,

b) $z = f(x, y(x)) = 2 - x - y(x)$, where $y(x) = -x + 1$, $|x| \le 1$,

c) $z = f(x, y(x)) = 2 - x - y(x)$, where $y(x) = -x^2 + 1$, $|x| \le 1$.

4. Describe or sketch the sets of points $\{(x, y, z)\}$ that satisfy

a) $z = f(x, y(x)) = (x^2 + y(x)^2)^{1/2}$, where $y(x) = x^2 + 1$, $|x| \le 1$,

b) $z = f(x, y(x)) = [1 - (x^2 + y(x)^2)]^{1/2}$, where $y(x) = x^3$, $|x| \le 1$

. . . .

The set of points $\{(x, y(x))\}$ or $\{(x, y): y = y(x), x \in I\}$ is the graph of a function $s: x \to y$ on I. And f may be considered as a function which maps the planar curve C_2 onto a curve C_3 in three-space, that is, $f(x, y(x))$ may be interpreted as a height z associated with the point $(x, y(x))$ (see Fig. 4.1). This is illustrated in

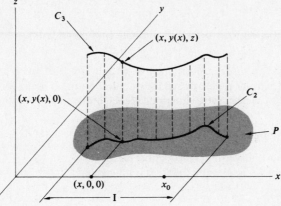

Figure 4.1

Exercises 3 and 4 [4.3.1]. If f is defined on a region P of the plane, and the curve $\{(x, y(x)): x \in I\}$ is in P, then the number $f(x, y(x))$ is defined for each pair $(x, y(x))$, where $x \in I$. That is, for $s \in \mathscr{F}$, a single value $f(x, y(x))$ is associated with each x in I. For x determines a single number $y(x)$, and x and $y(x)$ together determine a single value $f(x, y(x))$. The set of ordered pairs $\{(x, f(x, y(x)))\}$ is a function

$$\bar{f}: x \to f(x, y(x)), \qquad x \in I,$$

where $\bar{f}$ is a composite of the vector function (i, s) and f (Fig. 4.2).

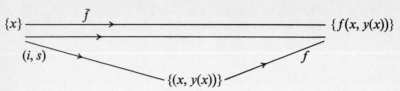

Figure 4.2

Let s on I be any solution function of $y' = f(x, y(x))$. Then $s \in \mathscr{F}$ and the corresponding $\bar{f}: x \to f(x, y(x))$ is well defined. Thus s satisfies $y' = \bar{f}(x)$. Hence, if $\bar{f}$ is continuous on I, then by Theorem 1.3.1 s may be specified as

$$y(x) = y_0 + \int_{x_0}^x \bar{f}(t)\, dt \quad \text{or} \quad y(x) = y_0 + \int_{x_0}^x f(t,\, y(t))\, dt, \quad x \in I.$$

EXERCISES [4.3.2]

. . . .

1. Consider the equation $y'(x) = 3xy^{1/3}$ and initial condition $y(0) = 0$. (See note in the proof of Theorem 1.13.1).

   a) Show that each of the functions $y_1(x) = x^3$ and $y_2(x) = 0$ on R is a solution.
   b) Specify the functions $\bar{f}_i$ determined by $y_1(x)$ and $y_2(x)$. Is each $\bar{f}_i$ continuous on R?
   c) Write the equivalent integral equation for each of the solutions defined by $y_1(x)$ and $y_2(x)$ on R.

2. Let $w(x) = [x]$, the greatest integer in x, $x \in [0, 3]$.

   a) Sketch the function $g(x) = \int_0^x w(t)\,dt$ on $[0, 3]$.
   b) Is w continuous? Is g continuous? Is g differentiable?
   c) Define a function $\hat{w}$ such that the integral function $\hat{g}$ is differentiable. What property is essential for $\hat{w}$?

3. Show that if $\bar{f}$ is continuous and s on I satisfies the integral equation

   $$y(x) = y_0 + \int_{x_0}^x \bar{f}(t)\, dt = y_0 + \int_{x_0}^x f(t,\, y(t))\, dt,$$

   then s satisfies the differential equation and initial condition. Indicate where the continuity of $\bar{f}$ is used in your argument.

4. a) If f is continuous on a region P, does that imply that $\bar{f}: x \to f(x, u(x))$ is continuous for any function u? Or does the function u have to be restricted? Give examples to illustrate your answers.
   b) Conjecture conditions on f and u that might ensure the continuity of $\bar{f}: x \to f(x,u(x))$.

. . . .

4.4 THE EQUIVALENCE OF $y' = f(x, y)$ AND THE INTEGRAL EQUATION
$y(x) = y_0 + \int_{x_0}^x f(t, y(t))dt$

To show that a solution of the integral equation

$$y(x) = y_0 + \int_{x_0}^x f(t, y(t)) \, dt$$

also satisfies the initial condition and differential equation, evaluate $y(x_0)$ and differentiate the integral expression (see Exercise 3 [4.3.2]). But is the integral expression differentiable? Statement S(1.3.2) indicates that the continuity of $\bar{f} \colon x \to f(x, y(x))$ is a sufficient condition. But while $\bar{f}$ is temporarily useful as an interpretation, it is not part of the given differential equation and hence cannot be restricted. The condition must be shifted to f and s. The following lemma suggests applicable conditions.

Lemma 4.4.1

If

a) f is continuous on a region P,
b) $s \colon x \to y$ is continuous on an interval I, and
c) $\{(x, y) \colon y = y(x), x \in I\} \subset P$,

then

$$\bar{f} \colon x \to f(x, y(x))$$

is continuous on I.

A proof is in Appendix D; Exercise 3 [4.4.1] includes suggestions for constructing a proof. Conditions (b) and (c) may be realized by restricting solutions to the subset $\mathscr{C}$ of continuous functions in $\mathscr{F}$. With this lemma, the equivalence of the integral equation may be stated as a corollary.

Corollary to Theorem 1.3.1

If

a) f is continuous on a region P of the plane,
b) $(x_0, y_0) \in P$,
c) $\mathscr{C}$ is the set of continuous functions whose graphs are in P,

then the system of equations, $y'(x) = f(x, y(x))$ and $y(x_0) = y_0$, is equivalent to the integral equation

$$y(x) = y_0 + \int_{x_0}^x f(t, y(t)) \, dt$$

relative to the set of functions $\mathscr{C}$.

EXERCISES [4.4.1]

· ·

1. Write a concise proof of the corollary above. [*Hint:* The arguments have already been given in the preceding section and exercises. Organize them to your own satisfaction.]

2. Does the restriction to set $\mathscr{C}$ eliminate any solutions of the differential equation? What properties do solution functions have? Might there be a solution in $\mathscr{F}$ which is not in $\mathscr{C}$?

*3. a) Consider intuitive arguments for Lemma 4.4.1. One is that a function continuous on a region ought to be continuous on any continuous path in the region. Another is that $\bar{f}$ is the composite, $f \circ (i, s)$, of two continuous functions.

* b) The conclusion of the lemma may be translated into ε, δ-terminology. Let x_1 be fixed and x_2 arbitrary in I. For every positive ε, there exists a number δ (depending on x_1 and ε) such that

$$|x_1 - x_2| < \delta \Rightarrow |f(x_1, y(x_1)) - f(x_2, y(x_2))| < \varepsilon.$$

The existence of δ may be shown by producing a number δ that works. Write the (a) and (b) parts of the hypothesis in ε, δ-terminology, using ε_1, δ_1 and ε_2, δ_2, respectively. Assume that the number ε is given, and try to determine a suitable δ in terms of δ_1 and δ_2.

· ·

4.5 PICARD'S SUCCESSIVE APPROXIMATIONS

If f is continuous on P, then relative to the set $\mathscr{C}$ of continuous functions whose graphs are in P, the differential equation $y'(x) = f(x, y(x))$, with initial condition $y(x_0) = y_0$, is equivalent to the integral equation

$$y(x) = y_0 + \int_{x_0}^{x} f(t, y(t)) \, dt.$$

But this integral expression does not specify a solution; as a definition, it is circular. And Exercise 1 [4.3.2] indicates there may be more than one solution. But the circularity suggests a method of defining functions that approximate solutions through (x_0, y_0). If, in the integrand, $y(t)$ is replaced by $s_0(t)$, where s is a specified function in $\mathscr{C}$, then the integral expression defines a specific function. For example, the constant function $s_0(x) = y_0$ defines a function

$$s_1(x) = y_0 + \int_{x_0}^{x} f(t, y_0) \, dt.$$

Consider the properties of such integral functions at (x_0, y_0). For the time being, assume that each function is defined on some interval I, no matter how small, containing x_0.

EXERCISES [4.5.1]

. . . .

1. a) Determine and compare the values $s_0(x_0)$, $s_0'(x_0)$, $s_1(x_0)$, $s_1'(x_0)$, $y(x_0)$, and $y'(x_0)$.
   b) For x close to x_0, is $s_0(x) = y_0$ or $s_1(x)$ closer to $y(x)$? Explain your answer.

2. a) Define a function s_2 by replacing $y(t)$ in the integrand by $s_1(t)$.
   b) Determine $s_2(x_0)$ and $s_2'(x_0)$.
   c) If you are familiar with the chain rule for partial differentiation, assume that f is differentiable and compare $y''(x_0)$, $s_1''(x_0)$, and $s_2''(x_0)$. *Hint:*

$$\frac{\partial f}{\partial y}(x_0, y_0) = \frac{\partial f}{\partial s}(x_0, s_0(x_0)).$$

3. a) Choose another simple function u_0, instead of s_0, and define the integral functions u_1 and u_2 as before. Compare the values of the u_i and their derivatives at x_0 with y_0 and $y'(x_0)$.
   b) Show that u_1 satisfies the initial condition, no matter what the choice of u_0.
   c) Show that u_2 satisfies the differential equation for any choice of u_0.

. . . .

The two curves $\{(x, s_1(x))\}$ and $\{(x, y(x))\}$ have the same tangent at (x_0, y_0). Thus, for x close to x_0, the value $s_1(x)$ is closer to $y(x)$ than is the constant y_0. If $s_1(t)$ replaces $s_0(t) = y_0$ in the integrand, one might expect the values of

$$s_2(x) = y_0 + \int_{x_0}^{x} f(t, s_1(t))\, dt$$

to be still closer to those of a solution function, in the vicinity of x_0. And s_2 defines an even better local approximation s_3, and so on. The functions $s_0(x) = y_0$ and

$$s_k(x) = y_0 + \int_{x_0}^{x} f(t, s_{k-1}(t))\, dt, \qquad k = 1, \ldots, n \ldots$$

are called *Picard's successive approximations*. The structure of the Picard functions is such that each successive function s_k seems to generate a better local approximation than the previous one. In order to examine an actual Picard sequence, consider a particular differential equation.

4.6 NUMERICAL AND POLYNOMIAL APPROXIMATIONS

EXERCISES [4.6.1]

. . . .

Consider the equation $y' = x - y$ with initial condition $y(0) = 1$.

1. Determine numerical approximations on the interval $[-2, +2]$. Use increments of 0.1 and estimate values to two decimals.
   a) Use the Euler method. Call this approximation S_{N1}.
   b) Use one of the modifications suggested in section 1.4. Call this approximation S_{N2}.

2. Sketch the numerical approximations carefully so that they may be used as a basis for comparison. (You may also wish to fit a solution curve on a direction field, but fine comparisons without numerical values are difficult.)

3. Define Picard's successive approximation functions s_k for $k = 0, 1, 2, 3$. Express these integral functions in terms of simple antiderivatives. Sketch the $s_k, k = 0, 1, 2, 3$, on the interval $[-2, +2]$.

4. Compare the Picard approximations to S_{Ni} and S_{N2} (Exercise 1). Determine the maximum difference $|s_k(x) - s_{Ni}(x)|$ for each pair (k, i), $k = 0, 1, 2, 3$; $i = 1, 2$; $x \in \{-2, -1.9, \ldots, 2\}$.

5. Define the s_k for $k = 4, 5, 6$ and obtain the polynomial functions in each case. Compare these approximations to S_{N1} and S_{N2} as in Exercise 4.

6. a) Continue to determine the s_k, $k > 6$, as polynomial functions until you are able to conjecture a limit function $s_\lambda(x) = \lim \langle s_n(x) \rangle$.
   b) Express s_λ as a finite combination of elementary functions.

7. a) What kind of equation is $y' = x - y$? Is it separable, or homogeneous, or linear? Use methods of Chapter 1 to obtain a solution function.
   b) Compare your solution with the limit function s_λ, conjectured in Exercise 6.

8. a) Compare curves for the solution function, the Picard approximations, and the S_{Ni}.
   b) Compare the values of the solution function and the s_k, $k = 0, 1, \ldots, 6$, at $x = \pm 0.5, \pm 1, \pm 2, \pm 10$.

9. Determine the maximum difference $|s_\lambda(x) - s_k(x)|$, for $k = 1, 2, \ldots, 6$, on the intervals $[-1, +1]$ and $[-2, +2]$ for $x \in \{-2, -1.9, \ldots, 2\}$. This difference is a measure of the accuracy of the approximation. Use this measure to compare the numerical approximations S_{Ni}.

10. Use R_n or the alternating series $s_\lambda(x) - s_k(x)$, $k = 1, 2, \ldots, 6$, to determine upper bounds on the errors in $s_k(x)$ on the intervals $[-0.5, +0.5]$, $[-1, +1]$, $[-2, +2]$, and $[-10, +10]$.

11. Compare the slopes $y'(x)$ and $s_k'(x)$ at $x = \pm 1, \pm 2, \pm 10$. Compare the first six derivatives at $(0, 1)$ of s_λ and the s_k, $k = 1, 2, \ldots, 6$.

. . . .

For the differential equation $y' = x - y$, the Picard approximating functions through $(0, 1)$ are simple polynomial functions (see Exercises 3 and 5 above). The approximations get successively better as k increases (see Exercises 4, 5, 6, 8, 9, 10, 11 above). Indeed, the sequence of functions $\langle s_k \rangle$ converges to a solution function (see Exercises 6 and 7 above).

4.7 UNIFORM CONVERGENCE AND THE
PICARD APPROXIMATION FUNCTIONS

Not all Picard approximation functions are as well behaved as those in Exercises 3, 5, 6, 7 [4.6.1]. But the convergence of the sequence $\langle s_k \rangle$ to a solution function s_λ is not just a stroke of good luck. It is almost inherent in the structure of the Picard functions

$$s_k(x) = y_0 + \int_{x_0}^x f(t, s_{k-1}(t)) \, dt.$$

For, if the sequence $\langle s_k \rangle$ converges, then the limit function

$$s_\lambda(x) = \lim \langle s_k(x) \rangle = \lim \langle s_{k-1}(x) \rangle.$$

This suggests that the integral equation defining $s_k(x)$ approaches

$$s_\lambda(x) = y_0 + \int_{x_0}^x f(t, s_\lambda(t)) \, dt.$$

Then s_λ satisfies the equivalent integral equation and hence, if continuous, is a solution of the differential equation.

To be more precise, this suggestion depends on the validity of a sequence of equalities:

$$s_\lambda(x) \overset{1}{=} \lim \langle s_k(x) \rangle \overset{2}{=} \left\langle y_0 + \int_{x_0}^x f(t, s_{k-1}(t)) \, dt \right\rangle$$

$$\overset{3}{=} y_0 + \lim \left\langle \int_{x_0}^x f(t, s_{k-1}(t)) \, dt \right\rangle$$

$$\overset{4}{=} y_0 + \int_{x_0}^x \lim \langle f(t, s_{k-1}(t)) \rangle \, dt$$

$$\overset{5}{=} y_0 + \int_{x_0}^x f(t, \lim \langle s_{k-1}(t) \rangle) \, dt$$

$$\overset{6}{=} y_0 + \int_{x_0}^x f(t, s_\lambda(t)) \, dt.$$

Equalities 1, 2, and 6 are simply definitions. Equality 3 is a basic property concerning the limit of the sum of two functions and is fundamental to the techniques of differentiation developed in elementary calculus (see Exercise 1 [4.7.1]). Equality 5 is an immediate consequence of the definition of a continuous function on a region P (see Exercise 2 [4.7.1]). Equality 4 is a fundamental question in analysis. Remember that the integral is itself a limit. So the question is one of commutativity. Can the order of two limits be interchanged? The answer no, not in all cases, is established by a single counterexample. Evaluate

$$\lim_{k \to 0} \lim_{h \to 0} \frac{h - k}{h + k} \quad \text{and} \quad \lim_{h \to 0} \lim_{k \to 0} \frac{h - k}{h + k}.$$

The question then is whether, under certain conditions, the order of limits may be changed. The problem appears in many guises. You may have met it previously relative to the mixed partials

$$\frac{\partial^2 f}{\partial x \partial y} \quad \text{and} \quad \frac{\partial^2 f}{\partial y \partial x}.$$

Specific and general theorems have been developed to clarify these situations. Equality 4 is valid for $x \in I$, if the Picard approximation functions converge uniformly on I. In this case, uniform convergence has a second equally important implication. If the functions s_k are continuous and uniformly convergent on I, then the limit function is continuous on I. Both of these theorems are assumed, but not proved here. (See Exercise 10 [3.3.1] and the discussion of uniform convergence in Section 3.7.)

EXERCISES [4.7.1]
. . . .

*1. State the limit property used in equality 3 on page 181. Write a proof of this property or look one up, if necessary.

*2. By definition, f is continuous at (x_0, y_0) if and only if

$$\lim_{(x,y) \to (x_0,y_0)} f(x, y) = f(x_0, y_0).$$

Assume f is continuous on P, and that the graph of s_λ on I, is in P. Show that for $t \in I$, $\lim \langle f(t, s_k(t)) \rangle = f(t, \lim \langle s_k(t) \rangle)$. List any other assumptions that are necessary.

. . . .

A Lipschitz condition on f assures the uniform convergence of the Picard approximation functions and, at the same time, the uniqueness of the solution function. This condition will be introduced inside the framework of the convergence arguments, where its form is intuitively acceptable. But before establishing the convergence, consider Picard integral functions for specific differential equations that expose difficulties which did not show up with the well-behaved function $f(x, y) = x - y$.

EXERCISES [4.7.2]

Consider the differential equations

a) $y' = e^x - y$ b) $y' = (1/x) - y$ c) $y' = x - 1/(y - 1)$

d) $y' = x - y/x$ e) $y' = (x + y)^{1/2}$ f) $y' = x - \ln x$.

1. Which of these equations do not have a solution through the point $(0, 1)$? Why not?

2. For each equation, indicate on the xy-plane the set of points $\{(x, y)\}$ at which the differential equation cannot be satisfied.

3. For each equation, choose a simple compatible initial condition.

4. Determine the Picard approximation functions s_0, s_1, s_2 through

     a) $(1, 0)$ for equations (b), (d), and (e),

     b) $(0, 1)$ for equations (a) and (f),

     c) $(0, 0)$ for equation (c).

5. For each s_k in Exercise 4, specify an interval $I \supset \{x_0\}$ on which s_k is defined and differtiable.

6. Sketch curves for each s_0 and s_1 for the initial condition in Exercise 4. Check each curve with the corresponding points indicated in Exercise 2. Compare each domain with the intervals chosen in Exercise 5.

7. For which equations does it seem reasonable to continue evaluating approximation functions in order to obtain a limit function?

8. Consider the domains consisting of the boundaries (in dashed lines in Fig. 4.3) and their interiors. In each case, assume that f is continuous on that region only, and indicate the maximum interval I on which s_1 is defined. Let $(x_0, y_0) = (1, \frac{1}{2})$.

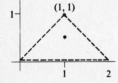
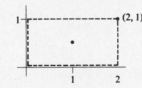
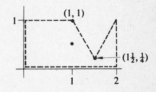

Figure 4.3

The exercises above raise difficulties. The integral expression for a Picard approximation function is easy to write, but the function may not be well defined unless the interval is restricted. (See Exercises 5 and 8 [4.7.2].) And even if the integral is well defined, it may be difficult or impossible to express s_k in terms of simple antiderivative functions. (See Exercise 7 [4.7.2].) Thus the Picard approximations do not provide a general method for writing a solution in simple form. In this sense, the existence theorem is less practical than those in Chapter 1. But the

fact remains that, in particular cases, the Picard method does specify the solution
in usable form or provides a tool for numerical approximations. That this limitation
obtains, even though convergence of Picard approximations to a solution function
can be assured in the general case, reflects a want of techniques. Evaluating the
limit of a sequence or series is usually far more difficult than determining the
convergence. Consider the number of tests you know for convergence, and
the number of techniques you have for evaluating limits.

The interval of definition for s_k is basic to the whole argument. If it is sometimes necessary to shorten the interval for successive functions s_k, is it possible to
determine a single interval on which every s_k is well defined? Without such a
common interval, the limit function s_λ would have no domain of definition. The
determination of this interval, the convergence arguments to follow, and certain
properties of s_λ all depend on the choice and manipulation of bounds on closed and
bounded sets.

4.8 BOUNDS AND CLOSED SETS

EXERCISE [4.8.1]

. . . .

1. Let $\langle a_n \rangle$ and $\langle b_n \rangle$ be convergent sequences on $(0, 1)$ and $[0, 1]$, respectively. Can the
   limits be outside the intervals? In each case, support your answer with examples or a
   conjecture and proof concerning limits of sequences on open or closed intervals.

2. Give examples of functions that are

   a) unbounded on a closed interval $[a, b]$, $a, b \in R$,
   b) unbounded and continuous on an open interval (a, b),
   c) unbounded and continuous on a closed interval $[a, b]$.
   If examples are impossible, make a conjecture concerning continuity and boundedness
   on certain types of intervals.

3. a) If f is continuous on $[a, b]$, indicate a bound for the integral $\int_a^b f(t)dt$. [*Hint:* Consider geometric areas. See Fig. 4.4.]

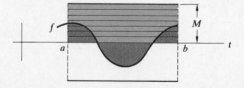

Figure 4.4

   b) If f is continuous on $I = \{x: |x - x_0| \le A\}$, find a bound for the integral $\int_{x_0}^x f(t)dt$,
   for all $x \in I$.

4. Consider a rectangular region

$$\{(x, y): |x - x_0| \le A, |y - y_0| \le B\} = P(A, B).$$

Define an unbounded continuous function on

a) $\{(x, y): |x - x_0| \le A, |y - y_0| < B\}$,
b) $P(A, B)$.

. . . .

If a sequence of points on a closed interval converges, the limit point is in the interval. (See Exercise 1 [4.8.1].) A continuous function on a bounded closed interval is bounded. (See Exercise 2 [4.8.1] and Appendix A.) This latter theorem suggests a method of obtaining bounds for integral functions. (See Exercise 3 [4.8.1].) In particular, $|\int_a^b f(t)dt| \le M|b - a|$ if M is a bound of f on $[a, b]$. These properties hold not only on bounded closed intervals, but also on certain sets in the plane or higher dimensions. For example, the rectangular region

$$\{(x, y): |x - x_0| \le A, |y - y_0| \le B\}$$

enjoys the same properties and is called a bounded closed region. In Euclidean space of any dimension, closed sets may be characterized by the first property stated above. Namely, a set is closed if, and only if, for every convergent sequence in the set, the limit point is also in the set. And a continuous function on a bounded closed set is bounded. These properties of closed sets, as those of uniformly convergent sequences and series, will be assumed and used, but not proved here. Why?

Ideally, basic assumptions, or axioms, should be simple, intuitively acceptable ideas to build on. Calculus has a disadvantage in that the basic concepts concern limits which, at first, are neither simple nor intuitively acceptable. Algebraic axioms, which describe operations on counting and rational numbers, offer a pleasant contrast. And theoretical work in algebra often paves the way for a careful treatment of the calculus in advanced analysis. But at this stage, it is probably wise to assume the properties of limits and continuous functions on bounded closed sets, after convincing yourself of their probable validity by trying enough examples to hazard conclusions in Exercises 1, 2, and 4 [4.8.1].

4.9 I_α, A COMMON DOMAIN FOR THE PICARD FUNCTIONS

Each s_k is defined on some interval I if s_{k-1} is in $\mathscr{C}$, the set of continuous functions whose graphs are in P. It is difficult to say more as long as P remains unspecified. But if P is a closed rectangular region,

$$P(A, B) = \{(x, y): |x - x_0| \le A, |y - y_0| \le B\},$$

then a single interval of definition can be specified for all the s_k. $P(A, B)$ separates the problems of limiting the domain of x and the range of $s_{k-1}(x)$.   Thus $f(x, s_{k-1}(x))\colon x \in I\}$ is in $P(A, B)$ if, and only if,

$$|x - x_0| \leq A \qquad \text{and} \qquad |s_{k-1}(x) - y_0| \leq B.$$

Since $P(A, B)$ is bounded and closed, f is bounded, that is, there is a number M such that $|f(x, y)| \leq M$ for all (x, y) in $P(A, B)$.

First consider s_1. The values $\{f(x, y_0)\}$ are well defined for $|x - x_0| \leq A$, and

$$|s_1(x) - y_0| = \left| \int_{x_0}^{x} f(t, y_0)\, dt \right| \leq M|x - x_0|.$$

If x is further restricted so that $|x - x_0| \leq B/M$, then

$$|s_1(x) - y_0| \leq M \cdot B/M = B.$$

(See Fig. 4.5.) Let α be the minimum value in $\{A, B/M\}$. Let

$$I_\alpha = \{x\colon |x - x_0| \leq \alpha\}.$$

Then, since

$$\{(x, s_1(x))\colon x \in I_\alpha\} \subset P(A, B),$$

s_2 is well defined on I_α.

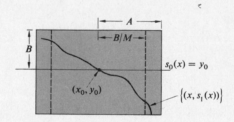

$$s_0(x) = y_0$$

$$(x_0, y_0)$$

$$\{(x, s_1(x))\}$$

Figure 4.5

EXERCISES [4.9.1]

1. a) Is s_3 defined on I_α? Justify your answer.
   b) Is s_3 continuous and/or differentiable? Why?

2. Is s_4 defined on I_α?

3. Show by induction that every s_k is defined on I_α.

4. Is every s_k continuous on I_α? Why?

5. a) Determine suitable rectangular regions $P(A, B)$ for the differential equations and initial conditions in Exercise 4 [4.7.2].
   b) Choose a bound M and determine I_α in each case.
   c) How should A, B, and M be chosen in order to maximize the domain I_α of the Picard approximation functions?

6. State a theorem concerning a common domain I_α for the Picard integral functions relative to the equation $y' = f(x, y)$.

7. Determine I_α for a function f, continuous on the strip $\{(x, y) \colon |x - x_0| \le A\}$.

8. Let f be continuous on the region P in Fig. 4.6. Intuitively, through which of the points (x_i, y_i) is there likely to be a solution curve? Which of the points is the center of a region $P(A, B)$ on which f is continuous?

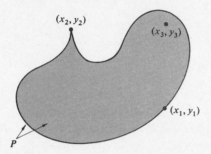

Figure 4.6

4.10 CONVERGENCE—THE LIPSCHITZ CONDITION

The concepts of convergence relative to a sequence or to a series are interchangeable. A series converges if, and only if, the sequence of partial sums

$$\left\langle S_n = \sum_{k=0}^{n} s_k \right\rangle$$

converges. And any sequence $\langle s_n \rangle$ represents the partial sums of a series

$$s_0 + (s_1 - s_0) + (s_2 - s_1) + \cdots + (s_k - s_{k-1}) + \cdots.$$

There seem to be more elementary convergence tests for series than for sequences. In particular, the comparison test offers a wide range of possibilities.

The sequence of Picard's successive approximations represents the partial sums of a series whose kth term is

$$\left(s_k(x) - s_{k-1}(x)\right) = \int_{x_0}^{x} f\big(t, s_{k-1}(t)\big) - f\big(t, s_{k-2}(t)\big)\, dt, \qquad x \in I_\alpha.$$

This term appears unwieldy. But if the integrand happens to be bounded by a multiple of $|s_{k-1}(t) - s_{k-2}(t)|$, then each term is bounded by a constant multiple of the preceding term. This suggests comparison with a geometric series. A Lipschitz condition provides precisely this situation: *f satisfies a Lipschitz condition on P* if, and only if, there exists a positive number L such that for any (t, y_i) in P,

$$|f(t, y_2) - f(t, y_1)| \le L|y_2 - y_1|.$$

Thus, if f is continuous and satisfies a Lipschitz condition on $P(A, B)$, then for $x \in I_\alpha$,

$$|s_k(x) - s_{k-1}(x)| = \left| \int_{x_0}^x f(t, s_{k-1}(t)) - f(t, s_{k-2}(t)) \, dt \right|$$

$$\leq \left| \int_{x_0}^x L|s_{k-1}(t) - s_{k-2}(t)| \, dt \right|.$$

Consider the series term by term on I_α. It has already been determined that $|s_1(x) - s_0(x)| \leq B$ or $M\alpha$ (see Section 4.9).

EXERCISES [4.10.1]

....

1. Use the Lipschitz condition to show that $|s_2(x) - s_1(x)| \leq ML\alpha^2$.
2. Determine a similar bound for $s_3(x) - s_2(x)$ in terms of L, M, and α.
3. Propose a bound for $s_k(x) - s_{k-1}(x)$. Establish your conjecture by induction.
4. Consider the series $\sum_{k=1}^\infty a_k$ where a_k is the bound of $|s_k(x) - s_{k-1}(x)|$. What type of series is $\sum_{k=1}^\infty a_k$? Does it converge on I_α? Does convergence impose any restrictions on α?
5. Does the sequence $\langle s_n \rangle$, where

$$s_n = s_0 + \sum_{k=1}^n (s_k(x) - s_{k-1}(x)), \qquad n \geq 1,$$

converge on I_α? Justify your answer.
6. Does $\langle s_n \rangle$ converge uniformly on I_α? Let

$$\lim \langle s_n \rangle = s_0 + \sum_{k=1}^\infty (s_k(x) - s_{k-1}(x)) = s_\lambda.$$

a) Show that

$$|s_\lambda(x) - s_n(x)| \leq M \frac{(L\alpha)^{n+1}}{L(1 - L\alpha)}.$$

b) If $L\alpha < 1$, can N be chosen so that $|s_\lambda(x) - s_n(x)| < \varepsilon$ on I_α for $n > N$?
7. State a theorem concerning the convergence of the Picard functions.
8. Does $f(x, y) = x - y$ satisfy a Lipschitz condition? What is the smallest possible number L? Compare the interval I_α, determined in Exercise 4 above, with the solution domain determined for $y' = x - y$ and $y(0) = 1$, in Exercise 7 [4.6.1].

....

The exercises above indicate that the Lipschitz condition implies the uniform convergence of the Picard integral functions. But the comparison series

$\sum_{k=1}^{\infty} (M/L)(\alpha L)^k$ converges only if $\alpha < 1/L$, so the common interval of convergence, I_α, depends on the Lipschitz number L. Exercise 8 suggests that L need not affect I_α. Perhaps more stringent bounds would yield a comparison series whose convergence is independent of L. Even though it is only necessary to produce a local solution, a restriction due to careless techniques is not desirable.

EXERCISES [4.10.2]
....

1. Let the bound of $s_1(x) - s_0(x)$ remain as $M|x - x_0|$.

   a) Show that

   $$|s_2(x) - s_1(x)| \leq LM \left| \int_{x_0}^x |t - x_0| \, dt \right|, \qquad x \in I_\alpha.$$

   b) Show that

   $$|s_2(x) - s_1(x)| \leq LM \int_{x_0}^x |t - x_0| \, dt | = \frac{M}{L} \frac{L^2(x - x_0)^2}{2}, \qquad x \in I_\alpha.$$

   [*Hint:* The antiderivative of the absolute value may be evaluated by considering the two cases, $x \geq x_0$ and $x \leq x_0$, separately, or by computing the area under the graph of $f(t) = |t - x_0|$ from x_0 to x.]

   c) Show that

   $$|s_3(x) - s_2(x)| \leq \frac{M}{L} \frac{L^3|x - x_0|^3}{3!}.$$

   d) Compare these bounds to the previous set of bounds (Exercises 1, 2 [4.10.1]).

2. Determine a similar bound for $s_4(x) - s_3(x)$.

3. Conjecture a bound for $s_k(x) - s_{k-1}(x)$.

4. Establish by induction the inequality

   $$|s_k(x) - s_{k-1}(x)| \leq \frac{M}{L} \frac{L^k|x - x_0|^k}{k!} \qquad \text{on} \quad I_\alpha.$$

5. a) What function has the following series expansion?

   $$\frac{M}{L} \sum_{k=0}^{\infty} \frac{L^k|x - x_0|^k}{k!}$$

   b) For what values of x does this series converge?

   c) On what interval does the sequence $\langle s_n \rangle$ converge?

6. Show that $\langle s_n \rangle$ converges uniformly on I_α. [*Hint:* Show that

   $$|s_\lambda(x) - s_n(x)| \leq \frac{M}{L} \sum_{k=n+1}^{\infty} \frac{(L\alpha)^k}{k!}$$

   for all x in I_α.]

7. Determine a measure of s_n as an approximating function.

a) Show that the remainder

$$|R_n(x)| = |s_\lambda(x) - s_n(x)| \leq \frac{M}{L} \frac{(L\alpha)^{n+1}}{(n+1)!} e^{L\alpha}.$$

b) Consider the differential equation $y' = x - y$ and initial condition $y(0) = 1$ (Section 4.6). Use the bound on R_n to estimate the accuracy of s_n, $n = 1, 2, \ldots, 6$, on the intervals $[-\frac{1}{2}, \frac{1}{2}]$, $[-1, 1]$, $[-2, 2]$, $[-10, 10]$. Compare these estimates with those in Exercises 9 and 10 [4.6.1].

8. State a theorem, concerning the convergence of the Picard approximation functions, with a stronger conclusion than that in Exercise 7 [4.10.1].

. . . .

A more careful choice of bounds shows that the sequence of Picard integral functions converges uniformly on I_α, $\alpha = \min\{A, B/M\}$ if f satisfies a Lipschitz condition on $P(A, B)$. The value of the Lipschitz number, L, does not restrict α.

4.11 s_λ IS A SOLUTION FUNCTION

Finally it must be shown that s_λ is continuous and satisfies the integral equation. In particular, the integral expression $\int_{x_0}^x f(t, s_\lambda(t))dt$ must be well defined. A sufficient condition is that s_λ be continuous and its graph be in $P(A, B)$.

1. s_λ *is continuous* since each s_n is continuous, and the convergence is uniform on I_α. An alternative proof, not using the uniform convergence argument, is suggested in Exercise 2 [4.11.1].

2. $\{(x, s_\lambda(x)): x \in I_\alpha\} \subset P(A, B)$. For any $\bar{x}$ in I_α, each number in the sequence $\langle s_n(\bar{x}) \rangle$ is in the interval $[y_0 - B, y_0 + B]$. Hence the limit point $s_\lambda(\bar{x})$ is also in the closed interval, and $|s_\lambda(x) - y_0| \leq B$. Since this holds for all x in I, the graph

$$\{(x, s_\lambda(x)): x \in I_\alpha\} \subset P(A, B).$$

(See Exercise 1 [4.11.1] concerning the position of $\{(x, s_\lambda(x)): x \in I_\alpha\}$ in $P(A, B)$.)

3. $\int_{x_0}^x f(t, s_\lambda(t))dt$ *is well defined* and differentiable. This is implied by results 1 and 2 above.

4. s_λ *satisfies the equation*

$$s_\lambda(x) = y_0 + \int_{x_0}^x f(t, s_\lambda(t)) \, dt$$

because of the uniform convergence and the continuity of $\bar{f} : x \to f(x, s_\lambda(x))$. (See Section 4.7.) An alternative proof is suggested in Exercise 3 [4.11.1]. Results 3 and 4 together imply that s_λ is differentiable.

Finally, since s_λ is continuous and satisfies the equivalent integral equation, s_λ is a solution of the equation $y' = f(x, y(x))$ and initial condition $y(x_0) = y_0$.

EXERCISES [4.11.1]

. . . .

1. The graph $\{(x, s_\lambda(x)): x \in I_\alpha\}$ is in a restricted portion of $P(A, B)$.

   a) Show that $|s_n(x) - y_0| \le M|x - x_0|$ for all n and all $x \in I_\alpha$.
   b) Sketch $|y - y_0| = M|x - x_0|$ on a geometric representation of $P(A, B)$, as shown in Fig. 4.7.

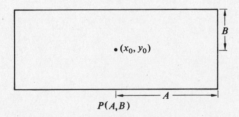

$$P(A,B)$$

Figure 4.7

   c) Sketch $|y - y_0| \le M|x - x_0|$ on $P(A, B)$.
   d) What is the implication of the inequality in (a) on the position of the graph $\{(x, s_\lambda(x)): x \in I_\alpha\}$ in $P(A, B)$?

2. Show that s_λ is continuous on I_α. [Alternative proof for result 1, Section 4.11.]

   a) Show that for every n and every x_1, x_2 in I_α,

   $$|s_n(x_1) - s_n(x_2)| \le M|x_1 - x_2|.$$

   [*Hint:* Express each $s_n(x_i)$ in integral form.]
   b) Use the argument about convergent sequences in closed intervals to show that, for all $x_i \in I_\alpha$,

   $$|s_\lambda(x_1) - s_\lambda(x_2)| \le M|x_1 - x_2|.$$

   c) Given $\varepsilon > 0$, find a number δ such that

   $$|x_1 - x_2| < \delta \Rightarrow |s_\lambda(x_1) - s_\lambda(x_2)| < \varepsilon.$$

3. Show that

$$\lim \left\langle \int_{x_0}^x f(t, s_n(t))\, dt \right\rangle = \int_{x_0}^x f(t, s_\lambda(t))\, dt \qquad \text{on} \quad I_\alpha.$$

   a) Show that

$$\left| \int_{x_0}^x f(t, s_\lambda(t))\, dt - \int_{x_0}^x f(t, s_n(t))\, dt \right| \le L|x - x_0|\, \frac{M(L\alpha)^{n+1}}{(n + 1)!}\, \frac{e^{L\alpha}}{L}.$$

(See Exercise 7 [4.10.2] and Section 4.7.)

b) Given $\varepsilon > 0$, determine an N such that

$$k > N \Rightarrow \left| \int_{x_0}^{x} f(t, s_\lambda(t)) \, dt - \int_{x_0}^{x} f(t, s_n(t)) \, dt \right| < \varepsilon.$$

4. Write an existence theorem for the differential equation $y' = f(x, y(x))$.

. . . .

4.12 THE SOLUTION s_λ ON I_α IS UNIQUE

Suppose s_λ and s_β are two solutions on I_α. Since both solutions satisfy the integral equation, both integral expressions are well defined on I_α. That is,

$$|s_\lambda(x) - s_\beta(x)| = \left| \int_{x_0}^{x} f(t, s_\lambda(t)) - f(t, s_\beta(t)) \, dt \right|.$$

Since f satisfies the Lipschitz condition,

$$|s_\lambda(x) - s_\beta(x)| \le L \left| \int_{x_0}^{x} |s_\lambda(t) - s_\beta(t)| \, dt \right| \le L\alpha \, 2B.$$

The procedure may be iterated; the integrand is the initial expression. Thus

$$|s_\lambda(x) - s_\beta(x)| \le L \left| \int_{x_0}^{x} L \left| \int_{x_0}^{t} |s_\lambda(u) - s_\beta(u)| \, du \right| dt \right|$$

$$\le L^2 \left| \int_{x_0}^{x} |t - x_0| \max_{x \in I_\alpha} |s_\lambda(t) - s_\beta(t)| \, dt \right|$$

$$\le L^2 \frac{(x - x_0)^2}{2} \max_{x \in I_\alpha} |s_\lambda(x) - s_\beta(x)| \le \frac{(L\alpha)^2}{2} \, 2B.$$

And the n-fold integral gives the bound

$$|s_\lambda(x) - s_\beta(x)| \le \frac{(L\alpha)^n}{n!} \, 2B, \qquad \text{for all} \quad n.$$

(See Exercises 1 and 2 below.) Since this set of bounds contains arbitrarily small numbers,

$$|s_\lambda(x) - s_\beta(x)| = 0 \qquad \text{and} \qquad s_\lambda(x) = s_\beta(x) \qquad \text{on} \quad I_\alpha.$$

EXERCISES [4.12.1]
. . . .

1. Show that

$$|s_\lambda(x) - s_\beta(x)| \le L^3 \frac{|x - x_0|^3}{3!} \max_{x \in I_\alpha} |s_\lambda(x) - s_\beta(x)|.$$

2. Use induction to establish that

$$|s_\lambda(x) - s_\beta(x)| \le L^n \frac{|x - x_0|^n}{n!} \max_{x \in I_\alpha} |s_\lambda(x) - s_\beta(x)|.$$

3. a) The Lipschitz condition ensures a unique solution on I_α. What does this imply about the function $f(x, y(x)) = 3xy^{1/3}$ on the region

$$P_2 = \{(x, y): |x - 0| \le 1, |y - 0| \le 1\}?$$

(See Exercise 1 [4.3.2].)

b) Given any positive number L, find two points in the region P_2 such that

$$|3xy_2^{1/3} - 3xy_1^{1/3}| \nleq L| y_2^{1/3} - y_1^{1/3}|.$$

c) Indicate a region on which f does satisfy the Lipschitz condition.

. . . .

4.13 A STATEMENT OF THE EXISTENCE AND UNIQUENESS THEOREM FOR $y' = f(x, y)$

Conditions ensuring the existence and uniqueness of a solution to the equations $y' = f(x, y)$ and $y(x_0) = y_0$ have been established. A solution has been constructed in a rectangular region centered at the initial point. The only task remaining is to write the theorem.

For an arbitrary equation $y' = f(x, y)$, the region P, on which f is both continuous and satisfies a Lipschitz condition, is not necessarily rectangular; the initial point may be anywhere in the region. The question is whether (x_0, y_0) is the center of a rectangular region wholly contained in P. If so, there is a unique solution on the corresponding I_α. Any point (x_0, y_0) which is the center of a rectangular region in P is called an *interior point* of P. The theorem applies only to interior points.

Theorem 4.13.1 [on solutions of $y' = f(x, y)$]

Let f be continuous and satisfy a Lipschitz condition on a region P of the plane. Let (x_0, y_0) be an interior point of P.

Then there exists a unique function on $I_\alpha \supset \{x_0\}$, satisfying the differential equation $y' = f(x, y)$ and the condition $y(x_0) = y_0$.

This solution may be defined as the limit of the sequence $\langle s_n \rangle$ on I_α, where

$$s_0(x) = y_0$$

and

$$s_n(x) = y_0 + \int_{x_0}^x f(t, s_{n-1}(t)) \, dt, \qquad n \ge 1,$$

M is a bound of f on

$$P(A, B) = \{(x, y): |x - x_0| \leq A, |y - y_0| \leq B\} \subset P,$$

and

$$I_\alpha = \{x: |x - x_0| \leq \alpha,$$

where $\alpha = \min \{A, B/M\}$.

4.14 THE PROOF OF THE EXISTENCE THEOREM FOR THE SECOND-ORDER LINEAR EQUATION

Theorem 2.1.1, the existence and uniqueness theorem for the second-order linear equation

$$y'' + p(x) y' + q(x) y = r(x), \tag{4.14.1}$$

has not yet been proved here. Are the methods, developed for the equation $y' = f(x, y)$, applicable to a second-order equation? The notation $y = y_1$ and $y' = y_2$ allows Eq. (4.14.1) to be written as a system of first-order equations,

$$\begin{cases} y_1' = y_2, \\ y_2' = -p(x) y_2 - q(x) y_1 + r(x). \end{cases} \tag{4.14.2}$$

Each equation in system (4.14.2) is of the form $y' = f(x, y)$, and a parallel procedure yields an existence theorem for the system of two equations. Consider each function f to be specified on a common three-dimensional domain,

$$\begin{cases} y_1' = f_1(x, y_1, y_2) = y_2 \\ y_2' = f_2(x, y_1, y_2) = -p(x) y_2 - q(x) y_1 + r(x). \end{cases} \tag{4.14.3}$$

The particular functions f_1 and f_2 are linear and continuous on any region $\{(x, y_1, y_2): x \in I\}$ if p, q, and r are continuous on I. (See Exercise 2 [4.14.1].) In this respect, each equation in (4.14.3) is simpler than the general equation $y' = f(x, y)$.

A proof parallel to that developed for Theorem 4.13.1 is outlined in the exercises; the reader may carry it through independently. Exercises [4.14.1] concern the equivalence of Eq. (4.14.1), the system (4.14.2), and a system of integral equations (all with initial conditions). And the Picard approximation functions are defined. Exercises [4.14.2] concern the Lipschitz condition relative to linear equations. Exercises [4.14.3] concern the convergence of the Picard approximation functions to a solution function. Exercises [4.14.4] concern the uniqueness of the solution function.

EXERCISES [4.14.1]

. . . .

[Assume that p, q, and r are continuous on I.]

1. a) Show that if $s: x \to y$ on I is a solution of Eq. (4.14.1), then y and y', written y_1 and y_2, satisfy the system (4.14.2).
    b) Conversely, if y_1 and y_2 satisfy (4.14.2), then $y_1 = y(x)$ is a solution of (4.14.1). Thus a solution of a second-order equation may be obtained by solving a system of first-order equations.

2. Determine a region P, in the three-space $\{(x, y_1, y_2)\}$, on which
    a) f_1 is continuous,
    b) f_2 is continuous,
    c) both f_1 and f_2 are continuous.

3. a) If $y_1(x)$ and $y_2(x)$ are defined on I, do the sets of pairs

    $$\{(x, f_i(x, y_1(x), y_2(x))\}$$

    define functions $\bar{f}_i$ on I for $i = 1, 2$?
    b) Write the $\bar{f}_i: x \to f_i(x, y_1(x), y_2(x))$ as composite functions.
★   c) Conjecture conditions which ensure the continuity of the f_i. (See Lemma 4.4.1.)
★   d) (For those who enjoy ε, δ-proofs.) Write a proof to establish your conjecture in (c).

4. State a condition which ensures that the integral function

    $$\int_{x_0}^{x} f_i(t, y_1(t), y_2(t)) \, dt$$

    is well defined for $x \in I$.

5. a) Write an equivalent integral equation for each equation

    $$y' = f_i(x, y_1(x), y_2(x)),$$

    with initial conditions $y_i(x_0) = y_{i,0}$, $i = 1, 2$.
    b) State the equivalence formally (see Corollary to Theorem 1.3.1 in Section 4.4).
    c) Establish the equivalence stated in (b). (See Section 4.4.)

6. Write the first three Picard approximation functions, $s_{i,0}$, $s_{i,1}$, $s_{i,2}$, $i = 1, 2$, for each integral equation.

7. Must the interval I be restricted so that each integral function is well defined? If so, how?

. . . .

Exercises [4.14.1] lead to conclusions parallel to those in the early part of this chapter. If p, q, r are continuous on I, then each f_i is continuous on

$$P_3 = \{(x, y_1, y_2): x \in I\}.$$

If $y_1(x)$ and $y_2(x)$ define continuous functions on I, then the set of pairs

$$\{(x, f_i(x, y_1(x), y_2(x)))\}$$

defines a continuous function on I for $i = 1, 2$. Hence for $x_0 \in I$, the integral expressions

$$\int_{x_0}^{x} f_i(t, y_1(t), y_2(t)) \, dt$$

are well defined on I. And each first-order equation $y_i' = f_i(x, y_1(x), y_2(x))$, with initial conditions $y_i(x_0) = y_{i,0}$, is equivalent to the integral equation

$$y_i = y_i(x_0) + \int_{x_0}^{x} f_i(t, y_1(t), y_2(t)) \, dt,$$

relative to the set of continuous functions on I. The Picard integral functions $s_{i,k}$ are well defined on P_3. Since the values $y_1(x)$ and $y_2(x)$ are arbitrary in P_3, the graph of each $s_{i,k-1}$ is in P_3 as long as x is in I. Hence, each $s_{i,k}$ is well defined on I. Thus I is a common domain for all the Picard integral functions.

Do the approximation functions $s_{i,k}$ converge uniformly? Theorem 2.1.1 does not hypothesize a Lipschitz condition. Do the functions f_i, perhaps because they are linear, satisfy a Lipschitz condition? (See Exercise 1 [4.14.2].) A function f satisfies a *Lipschitz condition* in y_i on P_3 if there is a positive number L such that

$$|f(x, y_1, y_2) - f(x, \bar{y}_1, \bar{y}_2)| \leq L \cdot (|y_1 - \bar{y}_1| + |y_2 - \bar{y}_2|)$$

for all such triplets in P_3.

EXERCISES [4.14.2]
. . . .

1. Consider the system (4.14.3).
    a) Do f_1 and f_2 satisfy a Lipschitz condition on some domain? Show that 1 is a Lipschitz number for f_1. Show that if $P \geq p(x)$ and $Q \geq q(x)$, $x \in I$, then $P + Q$ is a Lipschitz number for f_2 on I. Give another Lipschitz number for f_2.
    b) Is there a Lipschitz number that suffices for both functions?

2. Consider the third-order linear equation with continuous coefficients on I,

$$y''' + p_2(x)y'' + p_1(x)y' + p_0(x)y = r(x).$$

    a) Write a system of three first-order equations, whose solution determines a solution of this third-order equation.
    b) Define a Lipschitz condition applicable to a function on a domain in four-space.
    c) Do the functions, in this first-order system of linear equations, satisfy Lipschitz conditions on a common interval?

. . . .

In the system (4.14.3), both linear functions f_i satisfy a Lipschitz condition in y_1 and y_2 on a closed interval I (see Exercise 1 [4.14.2]). Exercise 2 suggests that a similar situation obtains for linear equations of higher order.

If it can be shown that the sequences $\langle s_{1,k} \rangle$ and $\langle s_{2,k} \rangle$ converge uniformly on I; that the limit functions $s_{1,\lambda}$ and $s_{2,\lambda}$ satisfy the integral equations; and finally, that $s_{1,\lambda}$ is a unique solution of the second-order differential equation with initial conditions, the goal will be accomplished. But even more can be accomplished with no extra effort. Suppose the convergence is established, assuming a Lipschitz condition, that is, disregarding the linearity or other simplicities of the special functions f_1 and f_2. Then the same arguments apply to other situations; for example, to a general, not necessarily linear, system of first-order equations, as well as to an nth-order linear equation.

Hence the convergence will be established for integral functions

$$s_{i,k} = y_{i,0} + \int_{x_0}^{x} f_i(t, y_1(t), y_2(t))\, dt,$$

which are assumed to be well defined on a common closed and bounded interval I, while the f_i are continuous and satisfy a Lipschitz condition on a closed region P. Let L_i be the Lipschitz number for f_i. Consider again the individual terms of the series

$$s_{i,0} + (s_{i,1} - s_{i,0}) + \cdots + (s_{i,k} - s_{i,k-1}) + \cdots .$$

EXERCISES [4.14.3]

1. a) Write the integral form of the terms $(s_{1,1} - s_{1,0})$ and $(s_{2,1} - s_{2,0})$.
   b) Show that for $x \in I$, closed and bounded, there is a number M such that

   $$|s_{1,1} - s_{1,0}| + |s_{2,1} - s_{2,0}| \le M|x - x_0|.$$

2. a) Find bounds for each $(s_{i,2} - s_{i,1})$ in terms of L_i, M and $|x - x_0|$.
   b) Show that

   $$|s_{1,2} - s_{1,1}| + |s_{2,2} - s_{2,1}| \le (L_1 + L_2) M \frac{|x - x_0|^2}{2}.$$

3. Show that

   $$|s_{1,3} - s_{1,2}| + |s_{2,3} - s_{2,2}| \le M(L_1 + L_2)^2 \frac{|x - x_0|^3}{3!}.$$

4. a) Conjecture a bound for $|s_{1,k} - s_{1,k-1}| + |s_{2,k} - s_{2,k-1}|$.
   b) Establish this bound by induction.
   c) Compare the series $\sum_{k=1}^{\infty} s_{1,k} - s_{1,k-1}$ to a positive numerical series.

5. State a conjecture concerning

   a) the convergence of $\langle s_{1,k} \rangle$ and $\langle s_{2,k} \rangle$.
   b) the continuity of the limit functions $s_{i,\lambda}$, on I, $i = 1, 2$.

6. a) Are the integral expressions $\int_{x_0}^{x} f_i(t, s_{1,\lambda}, s_{2,\lambda}) dt$ well defined?

   b) Do the $s_{i,\lambda}$ satisfy the integral equations

$$s_{i,\lambda} = s_{i,0} + \int_{x_0}^{x} f_i(t, s_{1,\lambda} s_{2,\lambda}) \, dt?$$

   c) What are the necessary conditions in each case? (See Section 4.11.)

. . . .

The sequences $\langle s_{1,k} \rangle$ and $\langle s_{2,k} \rangle$ converge uniformly on I to limit functions $s_{1,\lambda}$ and $s_{2,\lambda}$, respectively, which satisfy the corresponding integral equations. And if the integral equations are equivalent to the system of first-order equations, then $s_{1,\lambda}$ and $s_{2,\lambda}$ are solutions of these equations.

In the particular case of the system (4.14.2), $s_{1,\lambda}$ is a solution of the second-order equation (4.14.1). And the restriction that I be bounded and closed is unnecessary. If I is any interval and $x \in I$, there is a bounded closed interval I_c such that $x \in I_c \subset I$. A Lipschitz condition holds on I_c, since p and q are continuous on I_c and therefore bounded. Hence $s_{1,\lambda}$ is defined on I_c and, in particular, at x. Thus $s_{1,\lambda}$ is defined for all $x \in I$, even if $I = R$. (See Exercise 1 below.) A Lipschitz condition need not be hypothesized in Theorem 2.1.1.

So far, only the existence of a solution has been established. The uniqueness argument is also parallel; as before, the procedure of expressing the difference of two solution functions $s_{1,\lambda} - r_{1,\lambda}$ as an integral function and applying the Lipschitz conditions is iterated. (See Exercise 2 below.)

EXERCISES [4.14.4]

. . . .

1. Consider intervals which are not closed and bounded, such as $(a, b]$, $[a, b)$, $(-\infty, \infty)$, and (a, b). Show that for each interval I and any $x \in I$, there is a bounded closed interval I_c, such that $x \in I_c \subset I$.

2. Apply the methods used in Section 4.12 to establish the uniqueness of the solution $s_{1,\lambda}$ for the second-order linear equation with initial conditions.

3. Discuss an existence and uniqueness theorem for a third-order linear differential equation with initial conditions. Predict the similarities and differences in the procedure of proof. What sort of notation would simplify the writing of such a proof for order n?

4. Conjecture similar existence and uniqueness theorems and proofs for a system of first-order linear equations, as well as a system of first-order equations, not necessarily linear.

. . . .

Further information concerning existence theorems may be found in :

Coddington, Earl A., *An Introduction to Ordinary Differential Equations*, Englewood Cliffs, N.J.: Prentice Hall, 1961.

Greenspan, Donald, *Theory and Solution of Ordinary Differential Equations*, New York: Macmillan, 1960.

Numerical methods have been used throughout the book. Now that Taylor series, the remainder R_n, and the Picard integral approximations are familiar tools, it is possible to compare some of these methods.

4.15 NUMERICAL METHODS: RÉSUMÉ AND GENERAL DISCUSSION

Numerical methods for approximating solutions of differential equations stem from Taylor series, the sequence of Picard integral functions, or directly from the differential equations. The numerical calculations must, of course, be finite; the formulas are always some sort of polynomial approximation. Yet, it is the concept of the total series or sequence that offers a unifying view of the different methods, that allows estimates of the various errors involved, and that often leads to the development of new numerical procedures. In order to gain some perspective on present numerical methods in differential equations, the methods already presented will be classified and compared, extensions will be indicated, and some of the basic problems of numerical computation will be discussed relative to these methods.

Euler's method and its modifications (see Sections 1.4 and 4.6) are *one-step* methods. Starting with a single point (x_0, y_0) and a chosen distance $h = x_1 - x_0$, the value $y_1(x_1)$ is determined by calculating a slope directly from the first-order equation.

If $y' = f(x)$, $\qquad\qquad\qquad\qquad$ if $y' = f(x, y)$,

then by the Euler method:

$$y_1 = y_0 + hf(x_0); \qquad\qquad\qquad y_1 = y_0 + hf(x_0, y_0);$$

by a modified Euler method:

$$y_1 = y_0 + \frac{h}{2}\left[f(x_0) + f(x_1)\right] \qquad y_1 = y_0 + \frac{h}{2}\left[f(x_0, y_0) + f(x_1, y_1)\right]$$

The process is repeated so that each (x_i, y_i) is determined from values at (x_{i-1}, y_{i-1}). When $y' = f(x, y)$, the average slope of the modified method cannot be computed without y_1. A first estimate $y_1^{[1]}$, predicted by Euler's method, may be corrected by the modified method. Thus

$$y_1^{[2]} = y_0 + \frac{h}{2}\left[f(x_0, y_0) + f(x_1, y_1^{[1]})\right].$$

For greater accuracy, this predictor-corrector procedure may be iterated until $y_2^{[k+1]} = y_2^{[k]}$ within the desired accuracy; and so on at each succeeding step until $y_i^{[k+1]} = y_i^{[k]}$.

Multistep methods use values at two or more points (see Exercise 4 [2.5.1]). A simple multistep method is $y_{n+1} = y_{n-1} + 2h\, y_n'$. Thus values at (x_0, y_0) and (x_1, y_1) determine (x_2, y_2).

Before comparing methods, it should be emphasized that there are many numerical approximations for a given differential equation with initial conditions, even if there is a unique solution function. Each approximation depends on the particular method, the choice of step size h, and the number of decimal places used. A change in any one of these factors may change the approximation. The numbers x_0 and h determine a discrete domain for the approximation function. Function values for exponential, logarithmic, and trigonometric functions are provided by such approximation functions, commonly known as tables. Approximation functions provide useful models of physical situations. The accuracy of physical data is itself limited by the techniques of measurement. Numerical approximations may be chosen to provide the accuracy requisite in a particular situation. And often numerical approximations offer theoretical insights.

Step methods and predictor-corrector modifications are easily proliferated. Can the accuracy of these methods be compared without making the computations? The step methods above may be represented as Taylor polynomials, and the remainder R_n provides a measure of the accuracy (see Exercises [4.15.1]).

EXERCISES [4.15.1]

. . . .

[Euler methods and Taylor polynomials]

1. Give examples to show that step sizes h and $h/2$ may result in two different solutions which disagree even at common values of x.

2. Compute numerical approximations with $h = 0.10$ and two-decimal accuracy for the equations $y' = x - y$, $y(0) = 1$.

   a) Use a modified Euler method with an iterated predictor–corrector.

   b) Use a two-step method with a modified Euler method as a starter.

   c) Compare each solution with that obtained in Exercise 1 [4.6.1]. Use the measure of accuracy discussed in Exercise 9 [4.6.1].

3. a) Determine the first four Taylor coefficients, $y^n(x_0)/n!$, for series solutions of

   i) $y' = y$, $y(0) = 1$ on $[0, 1]$, $h = 0.10$,

   ii) $y' = e^x/x$, $y(1) = 0$ on $[0, 2]$, $h = 0.10$.

   b) Use the Lagrange form of the remainder R_n (Section 3.9) to get an upper bound for the truncation error. Use this bound to determine the number of significant decimals.

c) Compute the Taylor polynomial approximations. Compare the results with appropriate tables, or in (ii) with values obtained in Exercise 5(b) [1.4.1].

4. Show that the Euler method for $y' = f(x)$, $y(x_0) = y_0$, is equivalent to using the first two terms of a Taylor series for $y(x)$ at x_0. Estimate the error at each step in terms of the step size h.

5. Show that the modified Euler method, using the average slope, is equivalent to using the first three terms of a Taylor series. Compare the truncation-error bound with that of the Euler method (see Exercise 4).

6. a) Show that the two-step method $y_{n+1} = y_{n-1} + 2h\,y'_n$ is equivalent to a truncated series with error of order h^3. [*Hint*: Consider the difference of the Taylor series for $f(x_n + h)$ and $f(x_n - h)$ at x_0. Let $f(x_n \pm h) = y_{n+1}$. Compare the simplicity and accuracy of this method with the modified Euler method.]

7. If $y' = f(x, y)$, higher derivatives may be obtained by using the chain rule. For example,

$$y'' = \frac{d(y')}{dx} = \frac{\partial f}{\partial x}\frac{dx}{dx} + \frac{\partial f}{\partial y}\frac{dy}{dx}.$$

Obtain three- and four-term polynomial approximations to the solution of

$$y' = x - y,\ y(0) = 1.$$

· ·

Numerical methods are often composites of several methods. Among these composite methods, those of Runge-Kutta and Milne are relatively simple and widely used. Both use a weighted average of slopes and have a built in predictor-corrector procedure.

The fourth-order *Runge-Kutta method* is a single-step method. At each step, y is determined as $y_1 = y_0 + \frac{1}{6}(k_1 + 2k_2 + 2k_3 + k_4)$, where

$$\begin{aligned}
k_1 &= h\,f'(x_0, y_0),\\
k_2 &= h\,f'(x_0 + \tfrac{1}{2}h, y_0 + \tfrac{1}{2}k_1),\\
k_3 &= h\,f'(x_0 + \tfrac{1}{2}h, y_0 + \tfrac{1}{2}k_2),\\
k_4 &= h\,f'(x_0 + h, y_0 + k_3).
\end{aligned}$$

These formulas may be derived from the first five terms of a Taylor series. The derivation will not be given here, but Exercise 3 [4.15.2] indicates that the error is of magnitude h^5.

Milne's multistep method is specified by the formulas

$$y^{[1]}_{n+1} = y_{n-3} + \frac{4h}{3}(2y'_{n-2} - y'_{n-1} + 2y'_n),$$

$$y^{[2]}_{n+1} = y_{n-1} + \frac{h}{3}(y'_{n-1} + 4y'_n + y'_{n+1}).$$

At each step, if the error $E = y_{n+1}^{[1]} - y_{n+1}^{[2]}$ affects the decimals retained, h is decreased and the computations repeated. If E is relatively small, h may be increased.

EXERCISES [4.15.2]

[Runge-Kutta and Milne methods]

1. a) In the Runge-Kutta method, note that k_1 is the $y_1 - y_0$ determined by the Euler method. Identify each of the k_i in terms of a specific method and/or predicted and corrected values.
   b) In like manner, discuss the Milne formulas.

2. Compute numerical approximations for $y' = x - y$, $y(0) = 1$, with $h = 0.10$, using
   a) the Runge-Kutta method,
   b) the Milne method, with the Runge-Kutta method as a starter.
   c) Compare these approximations with those obtained in Exercises 1 [4.6.1] and 2 [4.15.1].

3. a) Show that if $y' = f(x)$, the Runge-Kutta method is Simpson's Rule, where the interval (x_0, x_1) is divided in two. [This interpretation suggests an estimate of the truncation error. The evaluation of the integral equation

$$y = y_0 + \int_{x_0}^{x} f(t)\, dt,$$

by Simpson's Rule, is equivalent to fitting a cubic polynomial to the curve

$$\{(x, f(x)), x\varepsilon[x_0, x_1]\}.$$

The antiderivative is a fourth-order polynomial and the error is of order h^5.]
   b) Is either of the Milne formulas related to Simpson's Rule?

The Runge-Kutta method, as a single-step method, is self-starting and the step size h may be varied easily. The Milne method, as a four-step method, needs a starter method for three of the initial values and for each subsequent change in h. Both methods are easily applicable to higher-order equations and systems of equations (see J. B. Scarborough, *Numerical Mathematical Analysis*, 6th edition, Baltimore: Johns Hopkins Press, 1966, pp. 312–318). Higher-order versions of the Runge-Kutta method give greater accuracy.

Step methods set up local procedures. New derivatives are calculated at each step. Each new function value is determined by previously computed values on a discrete subset. Thus each computed value in a step method incorporates the errors of all the previous computations.

By contrast, a truncated power series is valid to a certain degree of accuracy over an interval I. For example, the values $J_0(x)$ for $x \in [0, 2]$ (Exercise 1 [3.13.1]) may be computed independently with three-place accuracy using four terms of the series. Errors do not accumulate. The same accuracy may be obtained over a larger interval by using more terms of the series. The same situation obtains with a Picard approximation function; values are computed independently over an interval. For accuracy, these global methods of computation are ideal. Unfortunately, both methods have limited application. Series solutions with numerical coefficients can be obtained only for a restricted class of linear equations. If the series is defined by evaluating the Taylor coefficients directly from the differential equation, the higher derivatives may be difficult to evaluate. The alternative is to use only a few terms of the series, which in effect shortens the interval and reduces the procedure to a step method. In the Picard integral functions, as already noted (see Exercises [4.7.2] and discussion following), the integrand is sometimes difficult or impossible to integrate. A common procedure is to approximate the integrand by a polynomial, usually a difference formula. In this manner, the Picard integral function has become one of the chief sources of difference methods in the solution of differential equations. Such methods have been used for computing projectile motion with variable air density (see the end of the discussion in Section 1.7). Difference methods, which involve simple arithmetic procedures, are not considered here. They are easily available in books on numerical analysis. Roughly speaking, each step method is defined by difference formulas of one sort or another, and difference methods are local step methods.

The basic problem in numerical computation concerns the question of accuracy. What kind of errors are involved? How are they controlled and estimated? The *truncation error*, the only one discussed so far, is due to the approximating polynomial, that is, the reduction from a limit to a finite process. The computations themselves introduce a *roundoff error*. The computer, man or machine, uses a finite set of numbers having finite decimal representations. Each computation is rounded off to the nearest available number and thus introduces a small error. The truncation and roundoff errors cannot be avoided; neither are they independent. For example, a small step size decreases the truncation error, but increases the roundoff error since it increases the number of computations. Step size h should be chosen to minimize the combined error, but the optimal value is difficult to ascertain.

In global methods, the truncation error can be estimated, and the roundoff error is insignificant in independent computations. In step methods, there is sometimes a phenomenon called *instability*. Small errors early in the process may result in huge errors as the computations progress; indeed, the error may increase exponentially. This question of *stability* is actually a question of continuity. Is the output of the method, in the form of numerical values, a continuous function of the input? Or rephrased, can a small change in the data result in large differences in function values?

A specified degree of accuracy should be an essential part of each approximation. In practice, it may be difficult to estimate or measure the combined errors.

Numerical methods have been in use for centuries and have given rise to many developments in pure mathematics. The basic problems, other than the control and estimate of error, used to be eyestrain and time. Astronomers often spent years calculating orbits. Simple methods, as timesavers, were most desirable. High-speed computing machines have replaced these problems by one of expense. Complexity of program, high-storage requirements, and computing time all contribute to the cost.

The simple, accurate, all-purpose method does not exist. Each method has its strengths and weaknesses. For example, the Runge-Kutta method exhibits a high degree of accuracy and is stable, but it is expensive in computer time, and the errors are difficult to estimate. The Milne method is less expensive and truncation errors are easily estimated. However, the method is so unstable that it is not suitable for long runs on high-speed computers. It is often replaced by a similar four-step method of Adams-Bashford. The choice of method for a particular problem requires experience, intuition, and a certain measure of trial and error.

Computer science is in a state of rapid development. This science, too, is—and will probably always remain—an art. The potentialities of the art were indicated in the time and position predictions and corrections which guided the lunar missions in 1969 and 1970. To the mathematician, numerical methods offer approximate solutions to equations that cannot be solved by other methods, an approach to non-linear differential equations, and insights that often suggest new theories.

Further information concerning numerical methods in differential equations is found in:

Conte, S. D., *Elementary Numerical Analysis*, New York: McGraw-Hill, 1965.

Kelly, L. G., *Handbook of Numerical Methods and Applications*, Reading, Mass.: Addison-Wesley, 1967.

Scarborough, J. B., *Numerical Mathematical Analysis*, 6th ed., Baltimore: Johns Hopkins Press, 1966.

GENERAL READING SUGGESTIONS

Feynman, R. P., Leighton, R. B., and Sands, M., *The Feynman Lectures on Physics*, Vols. I and II: Reading, Mass.: Addison-Wesley, 1965.

Holton, G., *Introduction to Concepts and Theories in Physical Science*. Reading, Mass.: Addison-Wesley, 1952.

Hoyle, F., *The Black Cloud*, P37, Perennial Library. New York: Harper & Row, 1957 (paperback).

Park, D., *Contemporary Physics*, Chapter 1. New York: Harcourt, Brace, & World, 1964.

Scarborough, J. B., *Differential Equations and Applications*. Baltimore: Waverly Press, 1965.

1. BOUNDED SETS AND FUNCTIONS

A *set of real numbers $B = \{x\}$ is bounded* if there is a positive number M such that $|x| \le M$ for every $x \in B$.

A *function f is bounded on I*, if there is a positive number M such that $|f(x)| \le M$ for every $x \in I$. Or, rephrased, a function f is bounded on I if the set $f[I]$ is bounded.

2. CONTINUOUS FUNCTIONS

A function f is *continuous at x_0* if its domain includes an open interval containing

$$x_0 \quad \text{and} \quad \lim_{x \to x_0} f(x) = f(x_0).$$

one-sided
limits
$$\begin{cases} \lim_{x \to {}^+ x_0} f(x) = \lim_{h \to 0} f(x + h) & \text{where} \quad h > 0. \\ \lim_{x \to {}^- x_0} f(x) = \lim_{h \to 0} f(x + h) & \text{where} \quad h < 0. \end{cases}$$

A function f is *continuous from the left at x_0* if its domain includes

$$(x_2, x_0], \quad x_2 < x_0, \quad \text{and} \quad \lim_{x \to {}^- x_0} = f(x_0).$$

A function f is *continuous from the right at x_0* if its domain includes

$$[x_0, x_1), \quad x_0 < x_1, \quad \text{and} \quad \lim_{x \to {}^+ x_0} = f(x_0).$$

A function f is *continuous on an open interval I* if f is continuous at every $x \in I$.

A function f is *continuous on $[a, b]$* if f is continuous on (a, b) and is continuous from the right at a and from the left at b.

3. DIFFERENTIABLE FUNCTIONS

A function f is *differentiable at x_0*, if its domain includes an open interval containing x_0 and the number

$$f'(x_0) = \lim_{x \to x_0} \frac{f(x) - f(x_0)}{x - x_0}$$

exists.

A function f *has a right-handed derivative at* x_0 if its domain includes

$$[x_0, x_1), \, x_0 < x_1$$

and the number

$$f'(x_0) = \lim_{x \to {}^+x_0} \frac{f(x) - f(x_0)}{x - x_0}$$

exists.

A function f *has a left-handed derivative at* x_0 if its domain includes

$$(x_2, x_0], \, x_2 < x_0,$$

and the number

$$f'(x_0) = \lim_{x \to {}^-x_0} \frac{f(x) - f(x_0)}{x - x_0}$$

exists.

A function f *is differentiable on an open interval I* if f is differentiable at every $x \in I$.

A function f *is differentiable on* $[a, b]$, $a < b$, if f is differentiable on (a, b) and has a right-handed derivative at a and a left-handed derivative at b.

4. THE RIEMANN INTEGRAL $\int_a^b f(x)\, dx$

If f is a continuous function on $[a, b]$, $a < b$, the Riemann integral $\int_a^b f(x)\, dx$ is a number which can be defined as follows:

Let the finite set of numbers $\{x_i\}_n$, where $a_0 = x_0 < x_1 < \cdots < x_n = b$, be a partition P of the interval $[a, b]$.
Let $\bar{x}_i$ be a number such that $x_i \leq \bar{x}_i \leq x_{i+1}$.

The finite sum

$$\sum_{i=0}^{n-1} f(\bar{x}_i)(x_{i+1} - x_i)$$

is called a Riemann sum.

Let $|P|$ be the maximum value in the set

$$\{(x_{i+1} - x_i), \, i = 0, \ldots, n - 1\}$$

for the partition P.

Then

$$\lim_{|P| \to 0} \sum_{i=0}^{n-1} f(\bar{x}_i)(x_{i+1} - x_i) = \int_a^b f(x)\, dx.$$

An alternative definition uses upper and lower sums, U_p and L_p.

Let

$$U_p = \sum_{i=0}^{n-1} \overline{f(x_i)}(x_{i+1} - x_i),$$

where $\overline{f(x_i)}$ is the maximum value of f on $[x_i\ x_{i+1}]$, and

$$L_p = \sum_{i=0}^{n-1} \underline{f(x_i)}(x_{i+1} - x_i)$$

where $\underline{f(x_i)}$ is the minimum value of f on $[x_i, x_{i+1}]$.

Then $\int_a^b f(x)\,dx$ is the number such that

$$L_p \leq \int_a^b f(x)\,dx \leq U_p \qquad \text{for all} \quad P.$$

5. A DERIVATION OF THE RELATION $\dot{A} = \tfrac{1}{2}r^2\dot{\theta}$

Consider a curve $\{(r, \theta): r = r(t), \theta = \theta(t)\}$. The shaded area

$$A(t) - A(t_1) = A - A_1$$

is bounded by the areas of the sectors determined by the minimum and maximum values, r_m and r_M, of r on the interval $[t_1, t]$. For $\theta(t) \geq \theta(t_1)$,

$$\frac{\theta - \theta_1}{2\pi}\,\pi r_m^2 \leq A - A_1 \leq \frac{\theta - \theta_1}{2\pi}\,\pi r_M^2.$$

For $t > t_1$,

$$\frac{1}{2}\,r_m^2\,\frac{\theta - \theta_1}{t - t_1} \leq \frac{A - A_1}{t - t_1} \leq \frac{1}{2}\,r_M^2\,\frac{\theta - \theta_1}{t - t_1}.$$

The limits of each term as $t \to t_1$ are in the same relation, that is,

$$\tfrac{1}{2}r^2(t_1)\,\dot{\theta}(t_1) \leq \dot{A}(t_1) = \tfrac{1}{2}r^2(t_1)\,\dot{\theta}(t_1).$$

Thus $\dot{A} = \tfrac{1}{2}\,r^2\,\dot{\theta}$. (The same result holds for $\theta < \theta_1$ or $t < t_1$.)

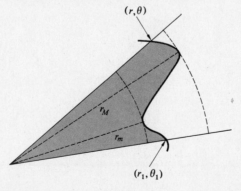

(r,θ)

r_M

r_m

(r_1, θ_1)

1. RADIAN MEASURE

Let θ denote the measure of an angle ABC. Consider a circle with center B and radius r. The radian measure θ is defined as the number s/r, where s is the arc length determined by angle ABC. Thus the straight angle has radian measure $\pi r/r = \pi$. Perpendicular rays with a common vertex define an angle of radian measure $\pi/2$.

Angle	Degree measure	Radian measure
	0	0
	30	$\pi/6$
	45	$\pi/4$
	60	$\pi/3$
	90	$\pi/2$
	180	π

The sine function is *not* a correspondence between angle and number, but a correspondence between two sets of numbers. The first set of numbers may, but need not, represent radian measures of angles.

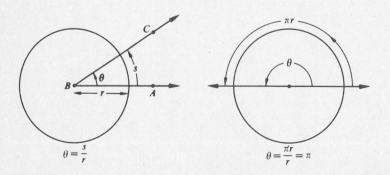

$$\theta = \frac{s}{r} \qquad\qquad \theta = \frac{\pi r}{r} = \pi$$

APPENDIX C

1. VALUES OF BESSEL FUNCTIONS J_0 AND J_1

TABLE 1

x	$J_0(x)$	$J_1(x)$
0	+1	0
1	+0.7652	+0.4401
2	+0.2239	+0.5767
3	−0.2601	+0.3391
4	−0.3971	−0.0660
5	−0.1776	−0.3276
6	+0.1506	−0.2767
7	+0.3001	−0.0047
8	+0.1717	+0.2346
9	−0.0903	+0.2453
10	−0.2459	+0.0435
11	−0.1712	−0.1768
12	+0.0477	−0.2234
13	+0.2069	−0.0703
14	+0.1711	+0.1334

2. ROOTS OF THE EQUATION $J_m(x) = 0$

TABLE 2

	$J_0 = 0$	$J_1 = 0$	$J_2 = 0$	$J_3 = 0$	$J_4 = 0$	$J_5 = 0$
1	2.405	3.832	5.135	6.379	7.586	8.780
2	5.520	7.016	8.417	9.760	11.064	12.339
3	8.654	10.173	11.620	13.017	14.373	15.700
4	11.792	13.323	14.796	16.224	17.616	18.982
5	14.931	16.470	17.960	19.410	20.827	22.220

1. PROOF OF LEMMA 4.4.1

Let

a) f be continuous on a region P,

b) $s: x \to y$ be continuous on an interval I,

c) $\{(x, y): y = y(x), x \in I\} \subset P$.

Then $\bar{f}: x \to f(x, y(x))$ is continuous on I.

Proof: Assume ε is given. By hypothesis (a), for any fixed (x_1, y_1) and arbitrary points (x_2, y_2) in P, there exists $\delta_1 > 0$, such that

$$|(x_2, y_2) - (x_1, y_1)| < \delta_1 \Rightarrow |f(x_2, y_2) - f(x_1, y_1)| < \varepsilon.$$

Considering the points (x_i, y_i) as vectors, we have

$$|(x_2, y_2) - (x_1, y_1)| = |(x_2 - x_1, y_2 - y_1)|$$
$$= [(x_2 - x_1)^2 + (y_2 - y_1)^2]^{1/2}.$$

Hence

$$|(x_2, y_2) - (x_1, y_1)| < \delta_1 \quad \text{if} \quad |x_2 - x_1| < \frac{\delta_1}{\sqrt{2}} \quad \text{and} \quad |y_2 - y_1| < \frac{\delta_1}{\sqrt{2}}.$$

Consider the particular fixed point $(x_1, y(x_1))$ and arbitrary points $(x_2, y(x_2))$. By hypothesis (b), for x_1 fixed and x_2 arbitrary in I, and the positive number $\delta_1/\sqrt{2}$, there exists $\delta_2 > 0$, such that

$$|x_2 - x_1| < \delta_2 \Rightarrow |y(x_2) - y(x_1)| < \frac{\delta_1}{\sqrt{2}}.$$

By hypothesis (c), $(x_1, y(x_1))$ and $(x_2, y(x_2))$ are in P for all $x_i \in I$. Thus if δ is chosen as the minimum value in the set $\{\delta_2, \delta_1/\sqrt{2}\}$,

$$|x_2 - x_1| < \delta \Rightarrow |f(x_2, y(x_2)) - f(x_1, y(x_1))| < \varepsilon. \quad \blacksquare$$

ANSWERS TO SELECTED PROBLEMS

[1.1.1]

1. $-f : x \to -f(x)$, thus $-f + f : x \to -f(x) + f(x) = 0$.

2. $1/f : x \to 1/f(x)$

3. If $f(x)$ is defined, so is $-f(x)$, but $1/f(x)$ is not defined if $f(x) = 0$.

4. Let F indicate the function graph, A the additive inverse, M the multiplicative inverse, I the inverse.

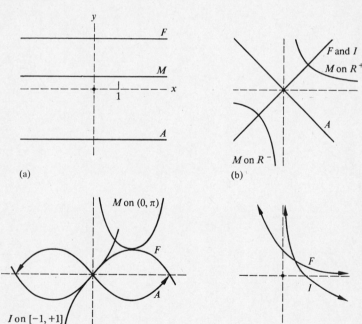

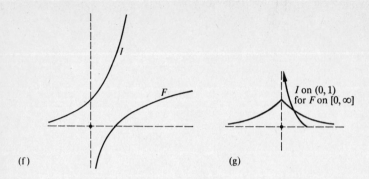

(f)

(g)

I on $(0, 1)$
for F on $[0, \infty]$

5.

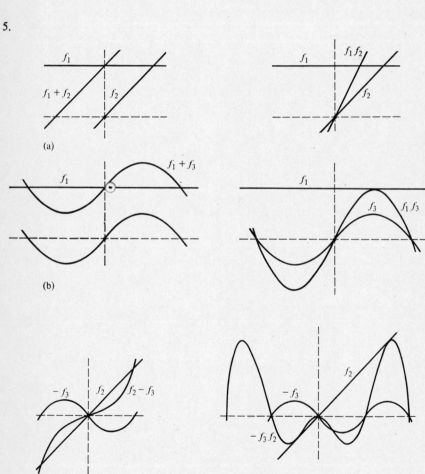

(a)

(b)

(c)

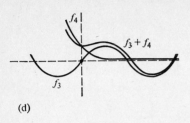 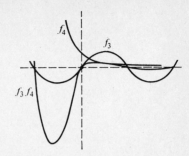

(d)

The graph of the sum function $f + g$ (or the product fg) may be sketched by plotting the sums $f(x) + g(x)$ [or the products $f(x)g(x)$] for particular values of x. Quick approximations may be made by using the additive and multiplicative properties of 0 and 1. For example, the product $f(x) \sin x$ is 0 whenever $\sin x = 0$, and is $\pm f(x)$ whenever $\sin x = \pm 1$. Intermediate values can be estimated by sight.

[1.1.2]

Let r_f indicate the constant function $r_f: x \to r, r \in R$.

1. $f''^2 + f' = 1_f$ 2. $y'^2 + 1 = 0$ 4. $xy' = e^x$

5. $2_f f' + f = 0_f$ 6. $y' = x^2 + 2x$ 7. $i^2 f'' + 2_f if' - 3_f f = i^3$

[1.2.1]

1.

2.

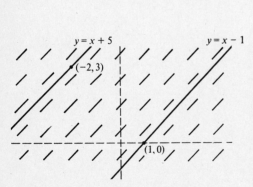

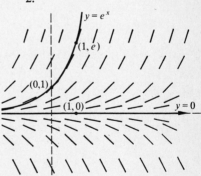

Solutions: lines with slope 1.

Solutions: exponential functions $(ke^x)' = ke^x$.

3.

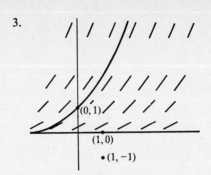

No solution on $(1, -1)$. A method for obtaining a solution through $(0, 1)$ is not obvious. Note that $y^{1/2}$ is undefined for $y < 0$.

4. *Hint:* At all points with a common x-coordinate, the slope y' is the same. Solution on $(0, 1)$:

$$y = \frac{x^3}{3} + 1.$$

Solution on $(1, 0)$:

$$y = \frac{x^3}{3} - \frac{1}{3}.$$

5.

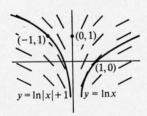

Solutions appear to be logarithmic. No solution on $(0, 1)$.

6. *Hint:* Note that $y' = y/x$ is the same for points $(1, \frac{1}{2})$, $(2, 1)$, $(3, \frac{3}{2})$, $(4, 2)$. Slope undefined on y-axis. Solution on $(1, 1)$: $y = x$, $x > 0$. Solution on $(1, 0)$: $y = 0$, $x > 0$. No solution on $(0, 0)$ because of the form of the differential equation; that is, $y' = y/x$ is undefined when x is 0.

7.

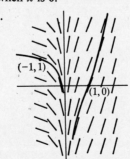

A method of obtaining solutions in terms of elementary functions is not obvious.

[1.3.1]

1. Both sets represent parabolas with vertices, $(0, c)$ or $(0, y_0 - \frac{1}{2}x_0^2)$, on the y-axis. The correspondence is not 1–1. Each number c corresponds to a set of pairs $\{(x_0, y_0)\}$, one pair for each point (x_0, y_0) on the parabola $y = \frac{1}{2}x^2 + c$.

2. a) $f(x) = x$ on R

   b) $y = \displaystyle\int_{x_0}^{x} u\, du + y_0 = \dfrac{x^2}{2} - \dfrac{x_0^2}{2} + y_0$

   c) $y_1 = x^2/2 - 2,\; y_2 = x^2/2 - 2$

5. a) $f(x) = 1/x$ on R^+ (or R^-)

   b) $y = \ln x - \ln x_0 + y_0$, on R^+ or $y = \ln x - \ln x_0 + y_0$, on R^-

6. a) $f(x) = e^x/x$ on R^+ (or R^-)

   b) $y = \displaystyle\int_{x_0}^{x} \dfrac{e^u}{u}\, du + y_0$

   c) $y_1 = \displaystyle\int_{1}^{x} \dfrac{e^u}{u}\, du + y_0$, on R^+, $y_2 = \displaystyle\int_{-1}^{x} \dfrac{e^u}{u}\, du + 1$, on R^-

8. a) $f(x) = \tan x$ on $(-\pi/2, \pi/2)$

   b) $y = \ln(\cos x_0/\cos x) + y_0$

   c) $y_1 = -\ln \cos x,\; x \in (-\pi/2, \pi/2),\; y_2 = 1 - \ln \cos x,\; x \in (-\pi/2, \pi/2)$. No solution through $(\pi/2, 1)$, $y_4 = -\ln |2\cos x|,\; x \in (\pi/2, 3\pi/2)$.

9. a) $f(x) = |\sin x|$ on R

   b) On $[0, \pi]$, $f(x) = \sin x$, $y = -\cos x + \cos x_0 + y_0$.   On $[\pi, 2\pi]$, $f(x) = -\sin x$, $y = \cos x - \cos x_0 + y_0$.

   c) $y_1 = 1 - \cos x,\; x \in [0, \pi],\; y_2 = \cos x,\; x \in [\pi, 2\pi]$. To extend y_1 on $[\pi, 2\pi]$, let $x_0 = \pi,\; y_0 = 2$; thus $y_1 = \cos x + 3$.

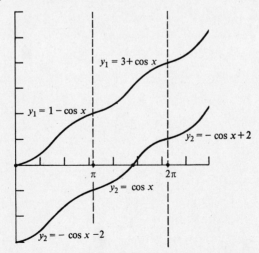

A solution curve on R may be obtained as on the graph above.

10.

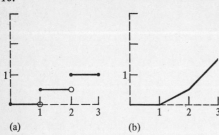

(a)   (b)   (c)   (d)

$A'(x)$ not defined at $x = 1$ or 2   There are many other functions satisfying (c) and (d).

[1.4.1]

1. b) $|y(x) - y_{approx}.| = 0.02$ at $x = 0.2$
   0.10 at $x = 1.0$
   0.20 at $x = 2.0$

2. a)

x	0.2	1.0	2.0
$y(x)$	-1.99	-1.55	0.10

3. $\frac{1}{2}(y'(x) + y'(x + h)) = x + h/2 = y'(x + h/2)$

5. a) $f(x) = e^x/x$ is continuous on R^+. Hence by Theorem 1.3.1, there is a solution through $(1, 0)$ on R^+. The solution is differentiable and $y' = e^x/x$. Because $e^x/x > 0$ on R^+, the solution is increasing. $y'' = (x - 1)e^x/x^2$. Therefore the curve is concave upward for $x > 1$, and downward for $x < 1$.

   b)

x	1	1.5	2.0
$y_{approx}.$	0	1.40	3.02
	0	1.41	3.01

   Euler method
   Modified Euler method (slope at midpoints)

6. Simpson's rule for approximating the definite integral,

$$\int_a^b f(x)\, dx$$

$$\doteq \frac{h}{3}\,[f(x_0) + 4f(x_1) + 2f(x_2) + 4f(x_3) + \cdots + 4f(x_{n-1}) + f(x_n)],$$

where $a = x_0$, $b = x_n$ and $h = x_i - x_{i-1}$, is found in most calculus texts.

[1.5.1]

1. $v = \int_0^t g\, du + v_0 = gt + v_0; h = \int_0^t (gu + v_0)\, du + h_0 = \frac{1}{2}gt^2 + v_0 t + h_0$

2. a) $h = -980t^2/2 + 1000t = 377.5$ cm at $t = \frac{1}{2}$, 510.0 cm at $t = 1$, 40 cm at $t = 2$

b) $v = 0 \Rightarrow t = 1.02$ and max $h = 510.2$ cm

c) Distance in 1 sec $= 510.0$ cm, in 2 sec $= 980.4$ cm

d) Same as above

3. a) h_{max} at $t = -v_0/g$

b) $h = h_0$ at $t = 0$ and $t = -2v_0/g$

[1.6.1]

Hint: The force changes every half time unit, hence the differential equation and initial conditions must be set up and solved on successive time intervals $[0, \frac{1}{2}]$, $[\frac{1}{2}, 1]$, $[1, \frac{3}{2}]$, etc.

1. a) $m = 1$, $v(0) = v_0 = 0$, $s(0) = s_0 = 0$

$$F = +1 \text{ on } [0, \tfrac{1}{2}]; \; ma = F \Leftrightarrow \dot{v} = +1 \Rightarrow v = \int_0^t 1 \, du + 0 = t; \; v(\tfrac{1}{2}) = \tfrac{1}{2}$$

$$s = \int_0^t u \, du + 0 = t^2/2; \; s(\tfrac{1}{2}) = \tfrac{1}{8}$$

$$F = -1 \text{ on } [\tfrac{1}{2}, 1], \; v_{1/2} = \tfrac{1}{2}, \; s_{1/2} = \tfrac{1}{8}; \; v = \int_{1/2}^t -1 \, du + \tfrac{1}{2} = -t + 1;$$

$$v(1) = 0$$

$$s = \int_{1/2}^t 1 - u \, du + \tfrac{1}{8}$$

$$= -(t^2/2) + t - \tfrac{1}{4}; \; s(1) = \tfrac{1}{4}$$

$F = +1$ on $[1, 1\tfrac{1}{2}]$, $v_1 = 0$, $s_1 = \tfrac{1}{4}$, and so on, for each half-minute time interval.

b) $s(10) = 2.5$ units

[1.7.1]

1. a) $\dot{v} = g - (k/m)v$. If $v_0 < mg/k$, then $0 < g - kv_0/m$ and the initial acceleration is positive, i.e., it opposes the fall of the body and exerts a braking action. (Remember that $g < 0 \Rightarrow mg/k < 0$, and hence $v_0 < 0$.) Thus the velocity increases (the magnitude or speed of fall toward the earth decreases) and approaches the terminal velocity from below.

b) $\dot{v}$ is positive if $v < mg/k$. Therefore v is increasing if $v < mg/k$. Thus $v^{\leftarrow}(v) = t$ exists on $(-\infty, mg/k)$.

c) $$v^{\leftarrow}(v) = t = \int_{v_0}^v \frac{1}{g - (k/m)u} \, du + 0 = (-m/k) \ln \frac{g - (k/m)v}{g - (k/m)v_0}$$

and $v = mg/k - (mg/k - v_0)e^{-(k/m)t}$ for $v < mg/k$

d) Unnecessary because $\dot{v} = 0$ and $v = c = mg/k$

2. a) If f is a 1–1 function *or* f is increasing (or decreasing) *or* $f' > 0$ (or < 0) on I, then $f^{\leftarrow}$ is defined on $f[I]$.

b) Intuitively, if a function curve is *smooth* on an interval, i.e., there are no breaks or sharp corners, then a slope is defined at each point and the function is differentiable. (One exception is an undefined "vertical slope"; hence the restriction in the hypothesis of the theorem that $f'(x) \neq 0$.)

c) Given that $f'(x) > 0$ on I, then $y = f(x)$ is an increasing differentiable function on I. At any $x_0 \in I$,

$$0 < \lim_{x \to x_0} \frac{y - y_0}{x - x_0} = \lim_{x \to x_0} \frac{1}{\frac{x - x_0}{y - y_0}} = \frac{1}{\lim\limits_{x \to x_0} \frac{x - x_0}{y - y_0}}.$$

And for an increasing function $x \to x_0 \Leftrightarrow y \to y_0$. Thus

$$\lim_{x \to x_0} \frac{x - x_0}{y - y_0} = \lim_{y \to y_0} \frac{x - x_0}{y - y_0}.$$

Note that this last property may not hold for functions that do not have inverses.

3. b) $v = -10 + (10 + v_0)e^{-t}$

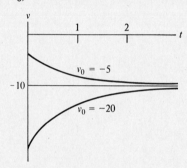

c) $h = -10t + (10 + v_0)(1 - e^{-t}) + h_0$

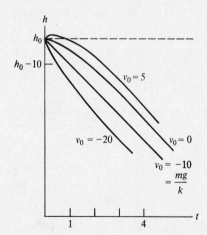

d) h_0 determines the starting point of the fall. Eventually the object falls at a rate close to the terminal velocity. If $v_0 < mg/k$, it falls rapidly at first and slows down to mg/k. If $0 > v_0 > mg/k$, it falls slowly at first and the speed increases toward $|mg/k|$.

4. $m = \frac{1}{2}$, $k = 0.6$, $g = -9.8$ m/sec $\Rightarrow mg/k \doteq -8$ m/sec, $v_0 = 0 \Rightarrow v = -8(1 - e^{-1.2t})$ and $h = -8t + 6.7(1 - e^{-1.2t}) + h_0$.

t	$e^{-1.2t}$	$1 - e^{-1.2t}$	v	$h_0 - h$
0	1	0	0	0
1	0.3	0.7	-5.6	3.4 m
2	—	—	—	—
3	—	—	—	17.6 m
4	—	—	—	—
5	—	0.998	-7.98	33.4 m

Stories	h_0	Approx. fall time	Approx. final velocity
5	15 m	3 sec	$v_c \doteq 7.7$ m/sec
10	30	5	$v_c \doteq 7.98$
100	300	—	$v_c \doteq 8$

5. a) 1 cu cm H_2O has mass 1 g, $m = (\frac{4}{3})\pi(0.05)^3 \doteq 0.5 \times 10^{-3}$; thus $mg/k \doteq -4$ m/sec
   b) $v \doteq -4(1 - e^{-2.5t})$
   $e^{-3} \doteq 0.05$; hence $|v - mg/k| < 5\%$ of 4 in about 1.2 sec
   $e^{-4.6} \doteq 0.01$; hence $|v - mg/k| < 1\%$ of 4 in less than 2 sec
   Approximate heights: 596 m and 592 m

6. Relative to a 1 mm raindrop, the mass of the hailstone is increased by a factor of 7^3. If k is increased by a factor of 7, then mg/k is 196 m/sec. Actually the fall behavior and terminal velocity of hailstones is complicated; experimental data are difficult to obtain and often contradictory. The degree of surface roughness, wetness, and temperature are all crucial, but not yet well understood. What evidence there is indicates that 196 m/sec is much too high; the simple model accounting only for a change in diameter is too crude. See J. T. Willis, K. A. Browning, and D. Atlas, "Radar observations of ice spheres in free fall," *J. Atmos. Sci.* **21**, 103–108, 1964.

7. $k \propto d$, $m \propto d^3$, $mg/k \propto d^2$; hence $10d \Rightarrow 100mg/k$, which is roughly true for the first three items in the table, and not thereafter.

8. On $[0, 3]$, $t_0 = 0$, $v_0 = 0$. Using $v(t) \doteq -60(1 - e^{-0.16t})$, $h - h_0 \doteq -60(t - 6.2(1.e^{-0.16t}))$. For $t \geq 3$, $t_0 = 3$, $v_0 \doteq -23$, $v(t) \doteq -6 - 17e^{-1.6(t-3)}$, $h - h_3 \doteq -6(t - 3) - 10[1 - e^{-1.6(t-3)}]$.

Approximate values

t	v	$h - h_0$	Fall/sec
1	-9	4.2	4.2
2	—	—	—
3	-23	39	19.0
		$h - h_3$	
4	-9.4	14	14
5	—	—	—
6	—	—	—
7	-6.0	34.0	6.1

After 7 sec the parachutist falls at about 6 m/sec.

[1.8.1]

1. b) $J = R$

c) $x = \int_0^y \dfrac{1}{1 + u^2}\, du + 0 = \arctan y$

   $y = \tan x$ on $(-\pi/2, \pi/2) = I$
   I cannot be extended; $\tan \pm \pi/2$ is undefined.
   d) $y = \tan (x - \pi/4)$ on $(-\pi/4, 3\pi/4)$
   $y = \tan (x + \pi/4)$ on $(-3\pi/4, \pi/4)$
   e) $(x_0 - \arctan y_0 - \pi/2, x_0 - \arctan y_0 + \pi/2)$

2. Theorem 1.8.1 is satisfied on $J = (0, \pi/2)$
   $e^x = \sin y$ or $y = \arcsin e^x$ on R^-
   b) Yes; $x = 0 \Rightarrow y = \arcsin 1 = \pi/2$
   $x > 0 \Rightarrow e^x > 1$ and $\sin e^x$ is undefined
   $x < 0 \Rightarrow e^x < 1$ and $\sin e^x$ is defined as $y \to 0$, $\sin y \to 0$ and $x \to -\infty$.
   Solution interval $I = (-\infty, 0)$

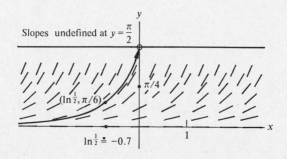

3. a) i) $y = e^{x-1}$ on R, $J = R^+$ ii) $y = -1/x - 2$ on $(-\infty, 2)$, $J = R^+$

$y = e^x$ on R $y = -1/x - 1$ on $(-\infty, 1)$, $J = R^+$

iii) $y = (2x - 1)^{1/2}$ on $(-\frac{1}{2}, \infty)$, iv) $y = (3x - 2)^{1/3}$ on $(\frac{2}{3}, \infty)$, $J = R^+$

$J = R^+$ $y = (3x + 1)^{1/3}$ on $(-\frac{1}{3}, \infty)$, $J = R^+$

$y = (2x + 1)^{1/2}$ on $(\frac{1}{2}, \infty)$,

$J = R^+$

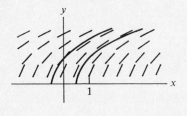
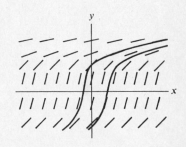

b) The domain of the functions may be extended in (iii) to $(\frac{1}{2}, 0)$ and $(-\frac{1}{2}, 0)$ and in (iv) to $x \le \frac{2}{3}$ and $x \le -\frac{1}{3}$.

c) The functions in (iii) and (iv) may include points $(x, 0)$. However, the functions will not be differentiable at points $(x, 0)$ in either (iii) or (iv).

[1.9.1]

1. a) No

b) Yes $(x_0, 0)$

c) $\lim_{v \to mg_k} (-m/k) \ln (g - (k/m)v)/(g - (k/m)v_0) = -\infty$ (no)

$\lim_{y \to 0} 2y^{1/2} - 2y_0^{1/2} = -2y_0^{1/2}$ (yes)

2. a) For every (x, y) on the line $y = b$, $y'(x) = 0 = q(b) = q(y)$. Hence $y = b$ is a solution of the differential equation $y'(x) = q(y)$.

b)
$$x = \int_{y_0}^{y} \frac{1}{q(u)} \, du + x_0$$

is a solution curve on (x_0, y_0) because the derivative is $1 = (1/q(y))y'$ or $y' = q(y)$. As $y \to b$,

$$x = \lim_{y \to b} \int_{y_0}^{y} \frac{1}{q(u)} \, du + x_0 = c + x_0,$$

and $\lim y' = \lim_{y \to b} q(y) = q(b) = 0$. That is, the slope of the solution curve approaches 0 as the points (x, y) on the curve approach $(c + x_0, b)$.

c) Thus the solution curve through (x_0, y_0) and the ray $\{(x, b): x \geq c + x_0\}$ together define a differentiable function which satisfies the equation $y' = q(y)$.

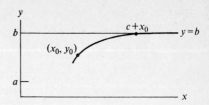

d)

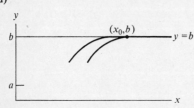

3. The solution curves, as $y \to b$, are asymptotic to, but do not meet the solution $y = b$. The interval J may be extended without losing the uniqueness of solutions. The equation $y' = y^2$ has solutions $y = 0$ on R, $y = 1/1 - x$ on $(-\infty, 1)$, $(x_0, y_0) = (0, 1)$, $y = -1/1 + x$ on $(-1, \infty)$, $(x_0, y_0) = (0, -1)$. No pair of solutions has a common point.

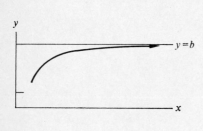

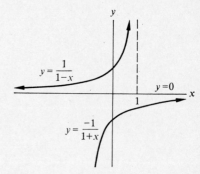

[1.11.1]

1. $\{(t, h): h > 0\}$; $q(h) \doteq -3(10^{-4})h^{1/2}$ is continuous and $\neq 0$ if $h > 0$.
$t = (\frac{2}{3})10^4 [h_0^{1/2} - h^{1/2}] + t_0, \ h > 0$
$h = [h_0^{1/2} - (\frac{3}{2})(10^{-4})t]^2$ for $t \in [t_0, \frac{2}{3}10^4 h_0^{1/2})$
Note that $h^{1/2}$ and hence h is not defined for $h < 0$, and because $h \leq 0$ the right half of the parabola,   , is not part of a solution.

2. In the region $\{(t, h): h \geq 0\}$ there may be many curves through one point. Both curves

$$h = \begin{cases} [\sqrt{40} - (\frac{3}{2})10^{-4}t]^2, & t \leq (\frac{4}{3})10^{4(1/2)} \\ 0, & t > (\frac{4}{3})10^{4(1/2)} \end{cases}$$

and

$$h = \begin{cases} [6 - (\frac{3}{2})10^{-4}t]^2, & t \leq (4)10^4 \\ 0, & t > (4)10^4 \end{cases}$$

go through the point $((\frac{4}{3})10^{4(1/2)}, 0)$.

4. The full tank empties in approximately 12 hours.
5. It takes almost 2 hours for the last centimeter to empty.
6. With uniform marks the difference in the first and last "hour" is about 3 of our usual hours. For uniform hours, the first marker is about 6.5 cm below the top and the last marker less than 0.3 cm above the bottom.

[1.12.1]

1. About 13,800 (± 100) B.C.,

$$t = \int_{A_0}^{A} -\frac{1}{ku}\, du + t_0 \qquad \text{or} \qquad t = \frac{-\ln A/A_0}{k}\;;$$

hence $k = (\ln 2)/5685$ and $t = (1/k) \ln \dot{A}_0/\dot{A}$.

2. This count suggests an accession date between 2200 and 2000 B.C. Note that both these questions were posed before 1955. The Stuiver-Seuss correction curve (see the discussion following these exercises) suggests an accession date closer to 2500 B.C. And the Lascaux paintings may be earlier than 16,000 B.C. See Colin Renfrew, "Carbon 14 and the Prehistory of Europe," *Scientific American* **225**, 4, 63–72, October 1971.

[1.13.1]

1. b) *Hint:* Formal differentiation gives $2f_1(x)f_1'(x) = 2x$.
   c) No
   d) The chain rule applies to a composite function $f_2 \circ f_1$ on I only if f_1 and f_2 are each differentiable on I and $f_2[I]$, respectively.
2. If $f_3(x) = -\sin x$ on R and $f_4(y) = (y - 1)^{1/2}$ on $[1, \infty)$, then $\int_0^x f_3(u)\, du \le 0$ for all x and $\int_1^y f_4(u)\, du > 0$ for $y > 1$. Hence these integral expressions do not define a function on any open interval containing $\{0\}$. There are many other such pairs of functions.

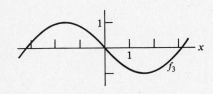 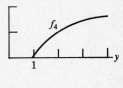

[1.13.2]

1. a) All points on the curve $yx = k$ have the same slope. Thus the slope is 1 at any point on the curve $xy = 1$ through points $(\frac{1}{2}, 2)$, $(1, 1)$, $(2, \frac{1}{2})$. This is often a convenient way to sketch direction fields.

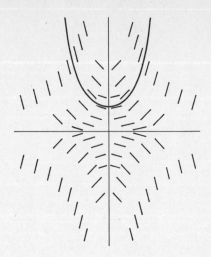

b) $I = R, J = R^+$

c) $\displaystyle\int_{x_0}^{x} u \; du = \int_{y_0}^{y} 1/u \; du$

On $(0, 1)$, $y = e^{x^2/2}$ on R. The solution $y = 0$ on $(0, 0)$ cannot be written in the usual form.

2. a)

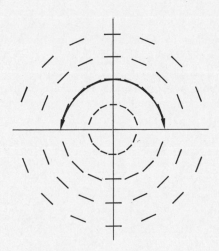

b) $I = R, J = R^+$

c) $x^2 + y^2 = 1$, $y > 0$, is a solution through $(0, 1)$ on $I = (-1, 1)$. No solution on $(0, 0)$. If J is extended to $[0, \infty)$, the solution $y = \sqrt{1 - x^2}$ is defined on $[-1, 1]$, but doesn't satisfy $y' = -x/y$ at $x = \pm 1$.

3. b) $I = R, J = R^+$
   c) On $(1, \frac{1}{16})$, $y = x^4/16$ on R. Thus the solution $y = 0$ through $(0, 0)$ is not unique.

4. b) $I = R^+, J = R^+$
   c) On $(1, 1)$, $y = x$ on R^+; on $(1, 0)$, $y = 0$ on R^+. No solutions through points $(0, y)$. If the region is extended to include $(0, 0)$, there is not a unique solution on $(0, 0)$.

5. a)   6. a)

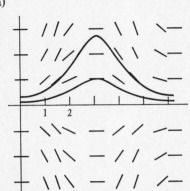
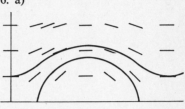

b) $I = R, J = R^+$ c) On $(\pi/2,$ b) $I = R, J = R^+$ c) On $(\pi/2, 1)$,
1), $y = e^{-\cos x}$ on R. On $(\pi, 1)$, $y = (1 - 2 \cos x)^{1/2}$ on $I = (\pi/3,$
$y = e^{-1-\cos x}$ on R. $y = 0$ is a $2\pi/3)$. On $(0, 1)$, $y = (3 - 2 \times$
unique solution on $(0, 0)$. $\cos x)^{1/2}$ on R.

d) 1–6 In Exercise 1 and 5, if J is extended to R, there are still unique solutions through every point in the plane. In Exercise 5 all solutions may be expressed in the usual form; not so in Exercise 1.

   If J is extended to include 0, there are unique solutions through every point in $\{(x, y): x \in R, y \geq 0\}$ in Exercise 6, and in Exercise 2 through every point except $(0, 0)$. In both cases, if J is extended to R, there are closed curves which do not define functions.

   If J is extended to include 0 in Exercises 3 and 4, there are no longer unique solutions through every point. J cannot be extended to R in Exercise 3 because the differential equation is not defined for $y < 0$. If I is extended to include 0 in Exercise 4, there is no longer a solution through every point, in particular through points $\{(0, y)\}$. Extended solution curves are not differentiable in Exercises 2 and 6.

7. On $J = (1, 2]$ $\ln y > 0$, and $-1 - \cos x \leq 0$ on R. Thus $\int_1^y u^{-1} \, du = \int_\pi^x \sin u \, du$ (or $\ln y = -\cos x - 1$) does not define a function. But $\ln y - \ln 2 < 0$ for $y < 2$. The condition that y_0 be interior is sufficient, but not necessary.

[1.14.1]

1. a) Approximately 11.0 km/sec or 6.96 mi/sec.
   b) To orbit 100 km above the earth, an object must travel with a speed of about

7.9 km/sec ($v \doteq 6400/\sqrt{102r_0}$; see Exercise 2[2.26.2] and answer). Therefore the exercise here is to show that, in order to have velocity 7.9 km/sec at a distance of 100 km, the object must be launched with velocity $v_0 \doteq 8.1$ km/sec. Then the object can be put into orbit by a suitable change in direction.

c) Approximately 11.2 km/sec. Since the propelling fuel is burning as a rocket leaves the earth and the escape velocity is attained after it leaves the earth's surface, resisting forces are relatively small.

3.
$$V = \left(V_0^2 - 2gr_E \left(\frac{r_E}{r} - 1 \right) \right)^{1/2}$$

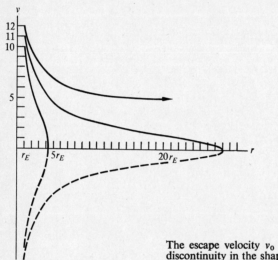

The escape velocity $v_0 = \sqrt{2|g|r_E}$ corresponds to a discontinuity in the shape of the curves. For smaller initial velocities, the object falls back to the earth.

[1.17.1]

1. The tangent of an angle determined by two intersecting lines with slopes m_1 and m_2 is given by

$$\tan \alpha = (m_2 - m_1)/(1 + m_1 m_2).$$

The differential equation is $yy'^2 + 2xy' - y = 0$.

[1.18.1]

1. The separable equation is

$$v' = -\frac{1}{x} (\sqrt{v^2 + 1}/v) [1 + \sqrt{v^2 + 1}].$$

Note that $v < 0$. Hence by definition $\sqrt{1/v^2} = -1/v$, a positive number.

2. $y^2 = [x_0(1 + \sqrt{(y_0^2/x_0^2) + 1})]^2 - 2x[x_0(1 + \sqrt{(y_0^2/x_0^2) + 1})]$, $x < 0$, $y > 0$, or y is the positive square root thereof. Note that $x_0 + \sqrt{y_0^2 + x_0^2} \neq x_0(1 + \sqrt{y_0^2/x_0^2 + 1})$.

3. $y^2 = -2K(x - K/2)$; S is the focus of the parabola.

4.
$$\lim_{x \to 0} y = |K|, \qquad \lim_{y \to 0} x = \tfrac{1}{2}K$$

5. $y = \sqrt{6(x + \tfrac{3}{2})}$, $-\tfrac{3}{2} \leq x \leq 0$ and $y \geq 0$

[1.19.1]

1. An integrating factor x results in $xy' + y = xe^x$. Then

$$xy = \int_{x_0}^{x} ue^u \, du + x_0 y_0$$

and the solution is

$$y = (1/x)e^x(x - 1) - e^{x_0}(x_0 - 1) + x_0 y_0.$$

2. Integrating factor $(1 + x^2)^{-1/2}$; solution $y = (x^2 + 1)^{-1/2}(x + 1)$

[2.1.1]

1. a) $y = \sin 2x$ b) $y = 0$ c) $y = \tfrac{1}{2} \sin 2x$ d) $y = 0$
   The solution in (a) is not unique.
   Straightforward methods for finding solutions will be developed in this chapter, but for the time being, use your ingenuity. Guess solutions, test them, and base successive guesses on the discrepancies you find.

[2.2.1]

2. $D_x g_1(x) + g_2(x) = D_x g_1(x) + D_x g_2(x)$ and $D_x kg(x) = k D_x g(x)$.

3. *Hint:* Carry through the same procedure as in Eq. (2.2.1). Use only the properties of the differential operator D and the definition of L.

[2.2.2]

1. a) A proof of Theorem 2.2.1. If f_1 and f_2, on I, are solutions of $L(y) = 0$, then $L(f_1) = 0$ on I, and $L(f_2) = 0$ on I. By the linearity of L, $L(c_1 f_1 + c_2 f_2) = L(c_1 f_1) + L(c_2 f_2) = c_1 L(f_1) + c_2 L(f_2) = 0$ on I. Hence $y = c_1 f_1 + c_2 f_2$, on I, is a solution of $L(y) = 0$.

[2.3.1]

2. $c_1(2 \sin^2 x) + c_2(1 - \cos^2 x) = c_1(2 \sin^2 x) + c_2(\sin^2 x) = 0$ if $c_1 = 1$ and $c_2 = -2$.

4.
$$c_1 f_1(x) + c_2 f_2(x) = 0 \begin{cases} \text{on } R^- \text{ if and only if } c_1 = -c_2, \\ \text{on } R^+ \text{ if and only if } c_1 = 0. \end{cases}$$

Hence there do not exist constants c_1 and c_2, not both zero, satisfying the relation $c_1 f_1(x) + c_2 f_2(x) = 0$ on R.

5. Do there exist constants c_1, c_2 (not both zero) such that $c_1 \sin x + c_2 \cos x = 0$ for $x = 0$ and $x = \pi/4$?

[2.4.1]

1. a) *Hint:* If the pair (x, y) satisfies the system of two equations, then $b_2(a_1 x + b_1 y - c_1) - b_1(a_2 x + b_2 y - c_2) = 0$. Use this relation to eliminate the variable y.

   b) and c) If

$$\begin{vmatrix} a_1 & b_1 \\ a_2 & b_2 \end{vmatrix} \neq 0,$$

then x and y are each defined as a ratio of two specific numbers. No other number pair could satisfy the simultaneous equations with these specific constants.

   d) i) There is a unique solution pair (x, y), and not both x and y are zero.
      ii) $(0, 0)$ is the unique solution.

   e) i) Consider the examples

$$\begin{cases} x + y = 1 \\ x + y = 0 \end{cases} \quad \text{and} \quad \begin{cases} x + y = 1 \\ x + y = 1 \end{cases}$$

   There may be no solution or many solution pairs.

      ii) Show that numbers x and y (not both zero) may be chosen to satisfy both equations in each of the cases

   i) $a_2 = b_2 = 0$ ii) $a_2 = 0, b_2 \neq 0$

   iii) $a_2 \neq 0, b_2 = 0$ iv) $a_2 \neq 0 \neq b_2$

   Only simple algebraic arguments are needed.

2. a) By examining the same cases considered in Exercises 1(e)(ii), you can show that the two lines are parallel if and only if the determinant is zero.

   The implications of these results are summarized in two statements that immediately follow this set of exercises. These statements are precisely the tools that are used to relate the Wronskian determinant to the independence of solution functions.

[2.4.2]

2. By hypothesis, $W(f_1, f_2) \not\equiv 0$ on I. If the differentiable functions f_1, f_2 are linearly dependent on I, then, by Theorem 2.4.1, $W(f_1, f_2) = 0$ on I. But this contradicts our hypothesis. Hence f_1 and f_2 must be linearly independent on I.

4. a) If $W(f_1, f_2) = (m - n)x^{n+m-1}$, then $W(f_1, f_2) \neq 0$ on R if $m \neq n$.
   b) $(x - 1)e^x \neq 0$ on R
   c) $1 - \ln x \neq 0$ on R^+

5. $W(f_1, f_2, f_3) = \begin{vmatrix} f_1(x) & f_2(x) & f_3(x) \\ f_1'(x) & f_2'(x) & f_3'(x) \\ f_1''(x) & f_2''(x) & f_3''(x) \end{vmatrix}, \qquad x \in I$

[2.4.3]

1. No. This would contradict Theorem 2.4.2.

2. a) Let f_1 and f_2, on I, be solution functions of an equation $L(y) = 0$. Then f_1 and f_2 are linearly dependent if and only if $W(f_1, f_2) \equiv 0$ on I.
   b) Let f_1 and f_2, on I, be solution functions of $L(y) = 0$. Then f_1 and f_2 are linearly independent if and only if $W(f_1, f_2) \neq 0$ on I.

[2.5.1]

3. a) $y = 3 \sin x$ b) $y = -\frac{1}{2} \sin x$
   c) $y = 0$ d) $y = \sin x + \cos x$
   e) $y = \cos x - \frac{1}{2} \sin x$

4. $y'' + y = 0$; $y(0) = 1$, $y'(0) = 1$, $y''(0) = -1$; $h = 0.2$. To start, $y''(0.1) = 1 + 0.1(-1) = 1 - 0.1 = 0.9$. Then the modified Euler method for a second-order equation is:

$$y(0 + 0.2) = 1 + 0.2(0.9) = 1.18,$$
$$y'(0.2 + 0.1) = 0.9 + 0.2(-1.18) = 0.664$$
$$y(0.2 + 0.2) = \qquad\qquad = 1.313,$$
$$y'(0.4 + 0.1) = 0.401.$$

Use 4 decimals for 3-decimal accuracy.

5. a) Maximum value of $f_1(x) = \sqrt{c_1^2 + c_2^2}$
   Maximum value of $f_2(x) = A$
   b) $f_2(x) = A[\cos \beta \sin x + \sin \beta \cos x]$

$$\cos \beta = c_1/\sqrt{c_1^2 + c_2^2}, \qquad \sin \beta = c_2/\sqrt{c_1^2 + c_2^2}; \qquad \beta \quad c_2$$
$$c_1$$

c) A is easily determined as in (a). The sine function is continuous. Hence for any number a such that $-1 \le a \le 1$, there is an angle of measure β_1 such that $\sin \beta = a$ and $\cos \beta = +\sqrt{1 - a^2}$ (or $-\sqrt{1 - a^2}$). Rephrased, given any two reals such that $a^2 + b^2 = 1$, there is an angle of measure β_1 such that $\sin \beta_1 = a$ and $\cos \beta_1 = b$. And $A \sin (x + \beta_1) = A \cos (x + (\beta_1 - \pi/2)) = A \cos (x + \beta_2)$.

d) i) This is a convenient form for the general solution and for imposing initial conditions $y(x_0) = y_0$ and $y'(x_0) = y_0'$.

   ii) This form indicates the amplitude and phase shift of the curve.

[2.6.1]

1. b) Base your implications on relations such as

$$\ddot{s} = -\frac{|g|}{l}\, s,$$

and $\dot{s} = 0$ when s attains a maximum or minimum value.

2. a) ii) $s = 5 \cos 2t + 7.5 \sin 2t$ (where s represents centimeters).

   c) π sec

   d) What change in t changes the argument from 0 to 2π? (That is, use the period of the sine and cosine functions.)

   e) Note that if $c_2 > 0$ then $0 < \beta < \pi$; that is, β is positive.

4. b) $A = \sqrt{s_0^2 + v_0^2 l/|g|}$

   c) $\beta = \arctan \left(\dfrac{s_0}{v_0} \sqrt{\dfrac{|g|}{l}} \right)$

6. 24.83 cm

7. Since the loss at the equator is small, do not use any approximation such as the length l, computed in Exercise 5. Rather express t in terms of the two gravitational values involved. The loss on the moon is somewhat larger. Interpret your answer; that is, clarify the meaning you give to the common phrase, "to lose time."

[2.9.1]

1. $F = m|g| = Km_Em/r_E^2$
   Using

$$|g| = Km_E/r_E^2 \doteq \begin{cases} 32 \text{ ft/sec}^2 \text{ and } r_E \doteq 4000 \text{ mi, or} \\ 9.8 \text{ m/sec}^2 \text{ and } r_E \doteq 6400 \text{ km,} \end{cases}$$

$$Km_E/r_E^3 \doteq \begin{cases} 10^{-4}/66 \text{ for distance in miles,} \\ \text{or } 10^{-4}/65 \text{ for distance in kilometers.} \end{cases}$$

2. Time to go through: approx. 42.5 min; period: approx. 85 min.
   Maximum speed at the center of the earth: a shade less than 5 mi/sec, or a shade more than 8 km/sec.

3. c) The period is the same in both tunnels, but the velocities differ.

[2.12.1]

1. a) $y = e^x$ and $y = e^{2x}$ are linearly independent.
   e) Solutions are linearly dependent on R.

4. Guess solutions for Exercise 1(e) and test them. Check your results with the method which follows the exercises.

[2.12.2]

1. *Hint:* First determine the coefficients p, q if the roots are m_1, m_1.

[2.12.3]

1. *Hint:* First determine the coefficients p, q if the roots of the auxiliary equation are $a \pm bi$. Then apply this particular operator to $f_1(x)$.

5. b) $y = e^{-x/2}(c_1 \cos (\sqrt{3}/2)x + c_2 \sin (\sqrt{3}/2)x)$.

7. a) $F = ma = -0.01v - (m|g|/l)s$ or $\ddot{s} + (0.01/m)\dot{s} + (|g|/l)s = 0$. The drag coefficient depends on the mass unit. (See note in Exercise 4 [1.7.1]). For simplicity, here assume that $k = 0.01$, where the mass unit is 1 g, and that the mass of the bob is 1 g.

   b) The particular constants are reasonably represented by the equation $\ddot{s} + 0.01\dot{s} + 4s = 0$, and the initial conditions indicate the solution $s = e^{-0.005t}(15 \cos 2t + 7.5 \sin 2t)$.

8. a) Using the same assumption stated in the answer to 7(a) above, $k/m = 1/1 = 1$.

9. a) $f'(t) = e^{at}(a \sin \omega t + \omega \cos \omega t) = e^{at} A(\sin \omega t + \beta)$, for suitable constants A and β. Hence the *critical values* of $y = e^{at} \sin (\omega t + \beta)$ occur for those values of t at which $\sin (\omega t + \beta) = 0$. Thus the maximum values occur at intervals of $2\pi/\omega$, as do the minimum values.

[2.13.1]

The constant H for a particular spring is determined by the displacement resulting from a single force.

$$H \Delta s = |F| \Rightarrow H(0.2) = 5 \Rightarrow H = 25.$$

1. $F = ma = F_H = -H_s \Rightarrow \ddot{s} + \dfrac{H_s}{m} = 0$

2. $\ddot{s} + 0.2\dot{s} + 250s = 0$, $s(t) = e^{-0.1t}[0.5 \cos 15.8t - 0.003 \sin 15.8t]$. In Exercises 2 and 3, if a computer and/or plotter are not available, sketch the additive terms and damping factor of the solution separately. Then add and multiply approximate function values at regular intervals. See Exercises [1.1.1].

4. Oscillations reach 0.01 of the original amplitude in 46 sec (Exercise 2) and 0.46 sec (Exercise 3).

6. $F_r = -3.16v$.

[2.17.1]

1. c) $y = e^{-x}(c_1 + c_2x) - 0.48 \sin 2x - 0.64 \cos 2x$.

   d) $y = e^{-(1/2)x}(c_1 \cos (\sqrt{3}/2)x + c_2 \sin (\sqrt{3}/2)x) + \frac{1}{3}e^x + x - 1$.

2. $r_1 = \tan x, r_2 = \ln x, \ldots$

3. $\ddot{s} + 250s = 20$, $sp = 0.08$, $s = 0.42(\cos 15.8t) + 0.08$.

5. $s = (c_1 \cos 15.8t + c_2 \sin 15.8t) + 0.067 \sin 10t$.

6. Does your final solution satisfy the initial conditions $s_0 = 0.5$, $v_0 = 0$? (The frequency $\sqrt{250 - 0.01}$ is reduced only slightly. Approximate as 15.8.)

7. $s = e^{-10t}(0.56 \cos 12.2t + 0.46 \sin 12.2t) - 0.063 \cos 15.8t$, or $= e^{-10t}(0.73 \sin 12.2t + 0.89) - 0.063 \cos 15.8t$.

9. a) In choosing a linear combination, there is no point in including any function which satisfies the associated homogeneous equation. For if $L_c(y_p) = 0$, then $y_p(x)$ contributes nothing to the term $r(x)$. The given linear combination satisfies $L_c(y) = 0$. Hence it is necessary to try more complicated functions, such as $t \cos 5\sqrt{10}\, t$ and $t \sin 5\sqrt{10}\, t$. These functions include the desired terms among their derivatives, but do not satisfy the associated homogeneous equation.

[2.18.1]

1. a) No maximum or minimum points: as $\omega \to \omega_0$, $P^2(\omega) \to \infty$

   b) $\ddot{s} + \omega_0^2 s = (A/m) \cos \omega_0 t$

2. b) $\omega = \sqrt{\omega_0^2 - \frac{1}{2}k^2/m^2}$. Establish that this frequency does in fact produce a *maximum* amplitude.

3. a) Approximate values for

	$k = 0.02$	0.2	2
Maximum amplitudes:	10	0.1	0.0017
Resonant frequencies:	15.81	15.75	7.1
4. Approximate half widths:	0.2	2	

For $k = 2$, the resonance curve does not reach half the maximum value on the left, but to the right of the maximum value, the half-width is about 10.

[2.24.1]

1. Suppose that $A''(x)/A(x) = K$, a positive constant. Then K can be denoted as $+\omega^2$ and the equation can be written as $A''(x) - \omega^2 A(x) = 0$. The auxiliary equation, $m^2 - \omega^2 = 0$, indicates the general solution $A(x) = c_1 e^{-\omega x} + c_2 e^{+\omega x}$. The boundary condition $A(0) = 0$ implies that $c_1 1 + c_2 1 = 0$; that is, $c_1 = -c_2$. And $A(l) = 0$ implies that $-c_2(e^{-\omega l} - e^{+\omega l}) = 0$. If $c_2 = 0$, then $A(x)$ is the zero function, which is of no interest as a model. If $c_2 \neq 0$, then $e^{-\omega l} = e^{+\omega l}$ or $e^{2\omega l} = 1$, which is impossible, since ω and l must be nonzero constants in any nontrivial situation. Hence $K \not> 0$. It is unnecessary to consider T and σ in this argument, because they are positive constants and do not affect the sign of the constant K.

2. The uniformly distributed mass, σL, is equivalent to mass m in the first model. The period of the fundamental vibration is that for $n = 1$, and in this second model it agrees with observation.

3. $y(x, t) = y_0 \cos t$, where $y_0 = y(x, 0)$ gives the initial position of the string.

7. The assumptions are stated or suggested in the discussion in Sections 2.21 and 2.22. But it is instructive to list them and to be well aware of all the possible variables and the restrictions that must be made to obtain the usual wave-equation model.

[2.25.1]

1. $n = 5 = 2^2 \cdot \frac{5}{4} \doteq 2^2 \cdot 2^{4/12}$, four halftones (or a third above the second octave).

[2.26.1]

1. a) $x = r \cos \theta$ and $y = r \sin \theta$, and r and θ are both functions of t. To determine $\ddot{x}$ and $\ddot{y}$, differentiate with respect to t and use the product and chain rules. Thus $\dot{x} = \dot{r} \cos \theta - r(\sin \theta)\dot{\theta}$.

   b) If the first equation is multiplied by $\sin \theta$ and the second by $\cos \theta$, subtraction eliminates the terms, including $-Gm_s$. The new set of equations are equivalent.

2. Integrating factor: r.

3. a) In differentiating $u(\theta) = 1/r$, remember that both sides must be differentiated with respect to the same variable. Thus $u'(\theta)\dot{\theta} = (-1/r^2)\dot{r}$.

   b) See Section 1.15, part 4 (pages 51 and 52) for a sketch and a discussion of the ellipse and the canonical form of the equation which displays the eccentricity as the coefficient of $\cos \theta$.

[2.26.2]

1. $e = 0$.

   The given equation represents an ellipse, and the coefficient of $\cos \theta$ is the eccentricity. [See Section 1.15, part 4 (pages 51–52).]

2. It is easier (and correct) to show that $v \doteq 6400/\sqrt{102r_0}$ km/sec or $4000/\sqrt{165r_0}$ mi/sec.

3. The same time as a round trip in an earth tunnel. (Note, from answer to Exercise 2, that $v \doteq 6400/\sqrt{102r_0}$, not $6400/\sqrt{165r_0}$ km/sec.)

4. Period about 105 minutes.

5. Approximately 500 km.

[2.27.1]

1. The complications include determinants of higher order and their evaluation requires more subtle techniques. The auxiliary equations are of higher degree, and the possible combinations of different types of roots increase in number. As the order of the differential equation increases, the method of undetermined coefficients becomes more cumbersome.

2. Roots of quadratic equations are easily obtained. There are also general, but more complicated, methods for solving third- and fourth-degree equations. But for algebraic equations of degree 5 and higher, there does not exist a general solution. Abel and Galois, who established these results, were brilliant mathematicians who died in their twenties. (See E. T. Bell, *Men of Mathematics*, Simon & Schuster, New York, 1937.) There are, however, special methods for solving particular types of equations, and numerical methods for approximating roots.

[3.2.1]

1. $\langle a_n \rangle = 1, 2, 3, \ldots, n, \ldots$
   $\langle b_n \rangle = 2.1, 4, 3, 6, 5, \ldots$ where $a_{2n-1} = 2n$ and $a_{2n} = 2^{n-1}$

2. b) 2, 4, 8, 16, 32
   c) $-1, 1, -1, 1, -1$
   f) $\frac{1}{2}, \frac{2}{3}, \frac{3}{4}, \frac{4}{5}, \frac{5}{6}$
   j) $-1, +\frac{1}{2}, -\frac{1}{3}, +\frac{1}{4}$
   l) 2, 2.25, 2.36, 2.44, 2.49 (to two decimals)
   m) 1, 0.5, $0.1\bar{6}$, $0.041\bar{6}$, $0.008\bar{3}$
   n) 2.000, 2.500, 2.667, 2.708, 2.716 (to three decimals)
   o) 4.00, 2.67, 3.47, 2.90, 3.34 (to two decimals)

3. Sequences (a) (e) (f) (g) (j) (m) converge to the limits 0, 0, 1, 0, 0, 0. For $\varepsilon = 0.1$, suitable n_ε are 4, 10, 10, 4, 10, 4. For $\varepsilon = 10^{-10}$, suitable n_ε are 4^{10}, 10^{10}, 10^{10}, 10^5, 10^{10}, 14. Any larger n_ε would suffice in each case.

4. Sequences (b), (c), (d), (h), (i), (k) diverge.

5. Without prior knowledge, l, n, o, must remain in the uncertain category. One of the difficulties in each of these cases is that it is not easy to conjecture a limit value L so that the definition of convergence may be applied.

6. a) $\frac{1}{2}$ c) 1 e) 1 f) 1 g) 1 i) 2 j) 1 k) 1 l) 3? m) 1 n) 2.8 o) 4
7. Sequences (b), (d), (f), (h), (l), (n) are nondecreasing.
8. a) Sequences (c), (i), (k) are bounded and divergent.
   b) Impossible
9. a) Yes; for example see sequences (a) and (e).
   c) No. Suppose $L_1 < L_2$ are both limits of a_n. Choose $\varepsilon < (L_2 - L_1)/2$. Is there an n_ε such that $|a_n - L_1| < \varepsilon$ and $|a_n - L_2| < \varepsilon$ for all a_n where $n > n_\varepsilon$?

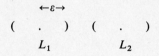

[3.2.2]

1. Sequences (f), (l), (n). [It may be difficult to establish these properties for sequences (l) and (n), but numerical values for $k = 1, \ldots, 6$ indicate that both properties probably hold.]
2. Sequences (a), (e), (g), (m) are bounded and nonincreasing and converge. Property $\overline{S.1}$: A nonincreasing bounded sequence of real numbers converges.
3. $3, 3.1, 3.14, 3.141$ (the sequence of decimal approximations to π) or $\pi - 0.1, \pi - 0.01, \pi - 0.001, \pi - 0.0001, \ldots, \pi - 10^{-n}, \ldots$. $l, n,$ and 0 are possibilities.
4. a) Let L be the least upper bound of a nondecreasing bounded sequence $\langle a_n \rangle$, that is, $a_n \le L$ for every n. For any $\varepsilon > 0$, $L - \varepsilon$ is not an upper bound of $\langle a_n \rangle$ since $L - \varepsilon < L$, the *least* upper bound. Therefore there is some a_{n*} such that $a_{n*} > L - \varepsilon$. $\langle a_n \rangle$ is nondecreasing; hence for all $n > n_*$, $a_{n*} \le a_n \le L$. Therefore $|L - a_n| < \varepsilon$ for $n \ge n_*$, that is, the definition of the convergence of $\langle a_n \rangle$ to L is satisfied.
   b) No. $3, 3.1, 3.14, 3.141, 3.1415, \ldots$ (the rational approximation to π).

$$\lim \left\langle \frac{n}{n+1}\,\pi \right\rangle = \pi, \qquad \lim \left\langle \frac{\pi}{n} \right\rangle = 0$$

5. a) Yes. $\lim \langle a_n + b_n \rangle = \lim \langle a_n \rangle + \lim \langle b_n \rangle = A + B$
   $|(a_n + b_n) - (A + B)| = |a_n - A + b_n - B| \le |a_n - A| + |b_n - B|$
   Given $\varepsilon > 0$, there is an $n_{\varepsilon/2}$ such that $|a_n - A| < \varepsilon/2$ and $|b_n - B| < \varepsilon/2$ for all $n > n_{\varepsilon/2}$. Hence for $n > n_{\varepsilon/2}$, $|(a_n - b_n) - (A + B)| < \varepsilon$.
   b) An argument similar to that in part (a) shows that $\lim \langle -b_n \rangle = B$ and $\lim \langle a_n - b_n \rangle = A - B$.
   c) For example,

$$\left\langle \frac{1}{2^n} + \frac{n}{n+1} \right\rangle$$

converges to 1.

6. Figure 3–1 indicates that the areas have the relationships

$$\ln(n+1) - \ln n > \frac{1}{n+1} \quad \text{and} \quad \tfrac{1}{2} + 1 + \cdots + \frac{1}{n-1} > \ln n.$$

Therefore

$$a_{n+1} - a_n = \frac{1}{n+1} - \ln(n+1) + \ln n < 0$$

and $a_n > 0$. Thus $\langle a_n \rangle$ is decreasing and bounded and hence convergent.

[3.3.1]

Note that the definition of $\lim \langle f_n \rangle = f_\lambda$ on I (just above the exercises) should read "$\lim \langle f_n \rangle = f_\lambda$ on I, if and only if, for each $x \in I$ and each $\varepsilon > 0$, there is an integer $n_{x,\varepsilon}$ such that for all $n > n_{x,\varepsilon}$, $|f_n(x) - f_\lambda(x)| < \varepsilon$."

1. a) If $x = 0 \lim \langle 0^n \rangle = 0$
   If $x = 1 \lim \langle 1^n \rangle = 1$

   b) If $x = -1 \langle (-1)^n \rangle$ diverges
   c) If $x \in (-1, 1)$, $\lim \langle x^n \rangle = 0$

2. a) If $\begin{cases} x = 0 \lim \langle n^0 \rangle = 1 \\ x = -1 \lim \langle n^{-1} \rangle = 0 \end{cases}$

   b) If $x = 1 \langle n^1 \rangle$ diverges
   c) Interval $[-1, -\tfrac{1}{2}]$

3. a) If $x = 0$ (or 1), $\lim \langle 1 + x^n \rangle = 1$ (or 2)

   b) If $x = 2$, $\langle 1 + x^n \rangle$ diverges
   c) $[0, 1]$

4. a) If $\begin{cases} x = 0, L = 1 \\ x = 1, L = \tfrac{1}{2} \end{cases}$

   b) If $x = -1$, odd terms not defined
   c) $[0, 1]$

5. a) If $x = 0$ (or 1) $\lim \left\langle \dfrac{x}{n} \right\rangle = 0$

   b) Converges for all x
   c) $(-\infty, +\infty)$

6. a) Diverges for all x

7. a) If $x = 0 \lim \langle e^{nx} \rangle = 1$
   If $x = -1 \lim \langle e^{-n} \rangle = 0$
   b) $x = 1$
   c) $[-2, -1]$

8. a) If $x = 0$ (or $\tfrac{1}{2}$) $\lim \langle (-x)^n \rangle = 0$
   b) $x = 1$
   c) $[\tfrac{1}{3}, \tfrac{1}{2}]$

9. a) If $x = \pi$ or 2π, $\lim \langle \sin nx \rangle = 0$
   b) $x = \pi/2$
   c) Impossible

10. In 1, on $[0, \tfrac{1}{2}]$, given $\varepsilon > 0$, any n_ε which suffices for $x = \tfrac{1}{2}$ will also work for all x, such that $x \geq 0$ and $x < \tfrac{1}{2}$. And similarly for 8 and the interval $[0, \tfrac{1}{2}]$.

11. b) $\{0, x: -1 < x < 1\} \cup (1, 1)$

c) $f_\lambda(x) = \begin{cases} 0 \text{ for } 0 \leq x < 1 \\ 1 \text{ for } x = 1 \end{cases}$

d) Each f_n is continuous on $[0, 1]$, but f_λ is not.

e) No. Given any $\varepsilon < 1$ and n_ε, there is an x close to 1 such that $|0 - x^n| > \varepsilon$ for some $n > n_\varepsilon$.

f) Yes; if N suffices for $x = \frac{1}{2}$, it suffices for all $0 < x < \frac{1}{2}$.

[3.4.1]

	$S_1, S_2, S_3, S_4, S_5, \ldots$	S_n
1. a) $1 - \frac{1}{2} + \frac{1}{4} - \frac{1}{8} + \frac{1}{16} - \frac{1}{32} \cdots$	$1, \frac{1}{2}, \frac{3}{4}, \frac{5}{8}, \frac{11}{16}, \frac{21}{32}, \ldots$	$\dfrac{?}{2^{n-1}}$
b) $0.1 + 0.01 + 0.001 + \cdots$	$0.1, 0.11, 0.111, 0.1111, \ldots$	$\underbrace{0.11111\ldots}_{n}100$
c) $0.3 + 0.03 + 0.003 + \cdots$	$0.3, 0.33, 0.333, \ldots$	$\underbrace{0.33\ldots}_{n}300$
d) $1 + 1 + 1 + 1 + 1 + 1$	$1, 2, 3, 4, 5, 6, \ldots$	n
e) $1 + \frac{1}{2} + \frac{1}{3} + \frac{1}{4} + \frac{1}{5} + \frac{1}{6} + \cdots$	$1, 1.5, 1.8\overline{3}, 2.08\overline{3}, 2.208\overline{3},$ $2.375, \ldots$	$?$
f) $-1 + 1 - 1 + 1 - 1 + 1 \cdots$	$-1, 0, -1, 0, -1, 0, \ldots$	$S_{2n} = 0,$ $S_{2n-1} = -1$
g) $1 - \frac{1}{2} + \frac{1}{3} - \frac{1}{4} + \frac{1}{5} - \frac{1}{6} \cdots$	$1, \frac{1}{2}, \frac{5}{6}, \frac{7}{12}, \frac{47}{60}, \frac{37}{60}, \ldots$	$?$
h) $1 + \frac{1}{2} + \frac{1}{6} + \frac{1}{24} + \frac{1}{120} + \cdots$	$1, 1.5, 1.6\overline{6}, 1.708\overline{3}, 1.7166,$ $1.718, \ldots$	$?$
i) $0.9 + 0.09 + 0.009 + \cdots$	$0.9, 0.99, 0.999, \ldots$	$\underbrace{0.999\ldots}_{n}900$

1. i) Converge (a) $S = ?$ Diverge (d)   Uncertain (e)
     (b) $S = 0.\overline{1}$ (f)   (g)
     (c) $S = 0.\overline{3}$
     (h) $S = ?$
     (i) $S = 0.\overline{9}$

2. a) $\displaystyle\sum_{k=1}^{\infty} \frac{1}{k^3} = \frac{1}{1} + \frac{1}{2^3} + \frac{1}{3^3} + \frac{1}{4^3} + \cdots$

$\displaystyle\sum_{k=2}^{\infty} \frac{1}{(k-1)^3} = \frac{1}{1} + \frac{1}{2^3} + \frac{1}{3^3} + \frac{1}{4^3} + \cdots$

$\displaystyle\sum_{k=-1}^{\infty} \frac{1}{(k+2)^3} = \frac{1}{1^3} + \frac{1}{2^3} + \cdots$

b) $\displaystyle\sum_{k=1}^{\infty} \frac{(-1)^{k+1}}{k^3}$, $\displaystyle\sum_{k=2}^{\infty} \frac{(-1)^k}{(k-1)^3}$, $\displaystyle\sum_{k=-1}^{\infty} \frac{(-1)^{k+1}}{(k+2)^3}$

(Note: exponents of (-1) may differ by a multiple of 2.)

c) $\displaystyle\sum_{k=0}^{\infty} (-\tfrac{1}{2})^k$, $\displaystyle\sum_{k=0}^{\infty} 10^{-k-1}$, $\displaystyle\sum_{k=0}^{\infty} 3(10)^{-k-1}$

3. a) $a_3 = \tfrac{1}{9} + \tfrac{1}{10} + \tfrac{1}{11} + \tfrac{1}{12} + \tfrac{1}{13} + \tfrac{1}{14} + \tfrac{1}{15} + \tfrac{1}{16}$

b) $a_0 = \tfrac{1}{2}, a_1 = \tfrac{1}{3} + \tfrac{1}{4} > \tfrac{1}{4} + \tfrac{1}{4} = \tfrac{1}{2}$, $a_2 = \tfrac{1}{5} + \tfrac{1}{6} + \tfrac{1}{7} + \tfrac{1}{8} > \tfrac{1}{8} + \tfrac{1}{8} + \tfrac{1}{8} + \tfrac{1}{8} = \tfrac{1}{2}$

Hence each $S_n > n/2$, and S_n diverges.

c) The harmonic series diverges.

4. a) $(1, \tfrac{1}{2})$: $1 + \tfrac{1}{2} + \tfrac{1}{4} + \cdots + \tfrac{1}{2}n + \cdots$,

$$S_n = \langle 1, \tfrac{3}{2}, \tfrac{7}{4}, \tfrac{15}{8}, \ldots, 2^n - 1/2^n - 1, \ldots \rangle$$

$(100, \tfrac{1}{2})$: $100 + 50 + 25 + \tfrac{25}{2} + \tfrac{25}{4} + \cdots$

$(0.1, -1)$: $0.1 - 0.1 + 0.1 + \cdots$

$(5, 0.01)$: $5 + 0.05 + 0.0005 + \cdots$ converges to $5.\overline{05}$

d) $S_n = a + ar + \cdots + ar^n$

$rS_n = \qquad ar + \cdots + ar^n + ar^{n+1}$

$$\overline{S_n(1-r) = a - ar^{n+1}}$$

Hence

$$S_n = \frac{a}{1-r}(1 - r^{n+1}).$$

e) $\displaystyle\lim \langle S_n \rangle = \frac{a}{1-r}$ if $|r| < 1$

5. a) $\displaystyle\frac{1}{2.3\ldots k} = \frac{1}{k!} \le \frac{1}{2^{k-1}} = \frac{1}{\underbrace{2.2\ldots 2}_{k-1}}$

b) Therefore $\displaystyle\sum_{k=1}^{n} \frac{1}{k!} \le \sum_{k=1}^{n} \frac{1}{2^{k-1}}$

c) $\sum_{k=1}^{\infty} 1/2^{k-1}$ is a convergent geometric series with sum 2.

d) *Hint:* The sequence

$$\left\langle \sum_{k=1}^{n} \frac{1}{k!} \right\rangle$$

is bounded by 2.

[3.5.1]

1. a) $0 \le 1/1 + 2^k < 1/2k$ and $\sum_{k=0}^{\infty} 1/2^k$ converges. Therefore by property $S - 2$, $\sum_{k=0}^{\infty} 1/1 + 2^k$ converges.

b) *Hint:* $0 \le (\sin^2 (1/2k))/k! \le 1/k!$.

c) Compare terms with $1/2^k$ or $1/k^2$.

2. $a_k \le b_k$ for all $k \Rightarrow \sum_{k=1}^{n} a_k \le \sum_{k=1}^{n} b_k$. Both

$$\left\langle \sum_{k=1}^{n} a_k \right\rangle \quad \text{and} \quad \left\langle \sum_{k=1}^{n} b_k \right\rangle$$

are increasing sequences, and each partial sum

$$\sum_{k=1}^{n} a_k \le \sum_{k=1}^{n} b_k \le B \le \lim \left\langle \sum_{k=1}^{n} b_k \right\rangle .$$

3. $\sum_{k=1}^{\infty} 1/2^k = \frac{1}{2} + \frac{1}{4} + \frac{1}{8} + \frac{1}{16} + \cdots$
   $\sum_{k=1}^{\infty} 1/k! = 1 + \frac{1}{2} + \frac{1}{6} + \frac{1}{24} + \cdots$
   For $k \le 3$, $1/k! > 1/2^k$. For $k \ge 4$, $1/k! < 1/2^k$. By comparison with $\sum_{k=4}^{\infty} 1/2^k$,
   $\sum_{k=4}^{\infty} 1/k!$ converges. Hence $\sum_{k=1}^{\infty} 1/k!$ converges, since

   $$\lim \left\langle 5/3 + \sum_{k=4}^{n} 1/k! \right\rangle = 5/3 + \lim \left\langle \sum_{k=4}^{n} 1/k! \right\rangle .$$

   The early terms of a sequence affect the sum, but not the convergence.

   $$\lim \left\langle \sum_{k=1}^{n} a_k \right\rangle = \lim \left\langle \sum_{k=6}^{n} a_k \right\rangle + \left\langle \sum_{k=1}^{5} a_k \right\rangle .$$

4. Property S.2. If, in the series $\sum_{k=0}^{\infty} a_k$, $a_k \ge 0$ for $k \ge N$, $\sum_{k=0}^{\infty} b_k$ converges, and $a_k \le b_k$ for $k \ge N$, then $\sum_{k=0}^{\infty} a_k$ converges.

5. If $a_k \ge b_k \ge 0$ for $k \ge N$ and $\sum_{k=0}^{\infty} b_k$ diverges, then $\sum_{k=0}^{\infty} a_k$ diverges.

6. b) $\sum_{k=1}^{n} qa_k = q \sum_{k=1}^{n} a_k$ and $\lim \langle q \sum_{k=1}^{n} a_k \rangle = qA$.
   c) *Hint:* Use the result in (b) with $q = -1$.

7. a) $S_2 = a_1 + a_2$ $S_1 = a_1$

   0

   $a_1 - a_2 + a_3 - a_4 = S_4$ $S_3 = a_1 - a_2 + a_3$

   b) $S_{2n} = S_{2n-2} + a_{2n-1} - a_{2n}$ and $a_{2n-1} - a_{2n} > 0$ since $\langle |a_n| \rangle$ is decreasing.
   d) $\mathrm{Lim} \langle S_{2n+1} - S_{2n} \rangle = \lim \langle |a_{n+1}| \rangle = 0$
   e) $\mathrm{Lim} \langle S_{2n+1} \rangle = \lim \langle S_{2n} \rangle$
   f) $|\sum_{k=n+1}^{\infty} (-1)^{k+1} a_k| = |S - \sum_{k=1}^{n} (-1)^{k+1} a_k| \le |a_n|$.
   (See diagram in (a).) Hence the error in approximating

   $$\sum_{k=1}^{\infty} (-1)^{k+1} a_k \quad \text{by} \quad \sum_{k=1}^{n} (-1)^{k+1} a_k$$

   is less than the last term used.
   g) $\langle 1 \rangle$ is neither decreasing nor has limit 0. From this evidence we cannot conclude that $\sum_{k=1}^{\infty} (-1)^k$ converges. $\langle 1/k \rangle$ is decreasing and has limit 0. Therefore $\sum_{k=1}^{\infty} (-1)^{k-1} 1/k$ converges.

8. a) $\sum_{k=2}^{n} 1/k^2 \le \int_{1}^{n} 1/t^2 \, dt = -1/t]_1^n = 1 - \dfrac{1}{n}$

   Therefore the increasing sequence $\langle \sum_{k=2}^{n} 1/k^2 \rangle$ is bounded by 1 and converges.

b) $1/k > \displaystyle\int_k^{k+1} 1/t \; dt$ and $\displaystyle\sum_{k=1}^{n} 1/k > \int_1^{n+1} 1/t \; dt = \ln(n+1),$

which is unbounded as n increases.

[3.5.2]

1. If $1 - \frac{1}{2} + \frac{1}{3} - \frac{1}{4} + \cdots (-1)^{k-1}1/k + \cdots = S$, then

$$+ \tfrac{1}{2} - \tfrac{1}{4} + \tfrac{1}{6} + \cdots = \tfrac{1}{2}S, \text{ and}$$

$$1 + \tfrac{1}{3} - \tfrac{1}{2} + \tfrac{1}{5} + \tfrac{1}{7} - \tfrac{1}{4} + \cdots = \tfrac{3}{2}S.$$

So the rearrangement of 2 positive terms followed by one negative term, increases the sum. Indeed, a rearrangement of the type

$$1 - \tfrac{1}{2} + \tfrac{1}{3} + \tfrac{1}{5} - \tfrac{1}{4} + \tfrac{1}{7} + \tfrac{1}{9} + \tfrac{1}{11} - \tfrac{1}{6}$$

$$> \tfrac{1}{2} \qquad\quad > \tfrac{1}{4} \qquad\qquad\quad > \tfrac{1}{6}$$

$$+ \tfrac{1}{13} + \tfrac{1}{15} + \tfrac{1}{17} + \tfrac{1}{19} + \tfrac{1}{21} + \tfrac{1}{23} - \tfrac{1}{8} + \tfrac{1}{25} + \cdots + \tfrac{1}{39} - \tfrac{1}{10} + \cdots$$

$$> \tfrac{1}{8} \qquad\qquad\qquad\qquad\qquad\qquad > \tfrac{1}{10}$$

results in a divergent series. On the other hand, a rearrangement of the absolutely convergent series

$$1 - \frac{1}{2} + \frac{1}{4} - \frac{1}{8} + \cdots + \frac{(-1)^{k-1}}{2^{k-1}}$$

does not change the sum.

2. a) $S_n^+ = \sum_{k=1, a_k > 0}^{n} a_k$ is the series consisting of all the positive terms of $\sum_{k=1}^{n} a_k$. Thus $S_n = S_n^+ + S_n^-$ and $S_n^+ - S_n^- = |S|_n$, $S_n^+ \le |S|_n \le |S|$, where $|S| = \lim \langle |S|_n \rangle$. Hence S_n^+ is an increasing and bounded sequence and converges if $|S|_n$ does. Similarly S_n^- is a decreasing series bounded from below by $-|S|$.
   b) $S_n = S_n^+ + S_n^-$ and $\lim \langle S_n \rangle = \lim \langle S_n^+ \rangle + \lim \langle S_n^- \rangle.$

3. Let $S = \lim \langle \sum_{k=1}^{n} a_k \rangle$, then $a_n = \sum_{k=1}^{n} a_k - \sum_{k=1}^{n-1} a_k$ and $\lim \langle a_n \rangle = \lim \langle \sum_{k=1}^{n} a_k \rangle - \lim \langle \sum_{k=1}^{n-1} a_k \rangle$. Therefore $\lim \langle a_n \rangle = S - S = 0$.

4. If $\lim \langle a_n \rangle = 0$, does $\sum_{k=1}^{\infty} a_k$ converge? Consider the harmonic series.

5. a) $\left.\begin{array}{l} |a_{n_0+1}| \le |a_{n_0}|q \\[4pt] |a_{n_0+2}| \le |a_{n_0+1}|q \le |a_{n_0}|q^2 \\[4pt] |a_{n_0+3}| \le |a_{n_0+2}|q \le |a_{n_0}|q^3 \end{array}\right\} \Rightarrow \begin{cases} \sum_{k=n_0}^{\infty} a_k \text{ converges absolutely.} \\ \text{by comparison with the geometric series} \\ \sum_{k=0}^{\infty} |a_{n_0}|q^k. \\ \text{Hence } \sum_{k=0}^{\infty} a_k \text{ converges.} \end{cases}$

   b) There is a number p such that $q < p < 1$, and for $\varepsilon = p - q$ there is an integer n_0 such that for $n > n_0$,

$$\left| \left| \frac{a_{n+1}}{a_n} \right| - q \right| \le \varepsilon.$$

Hence

$$\left| \frac{a_{n+1}}{a_n} \right| \le p < 1$$

for all $n > n_0$ and $\sum_{k=1}^{\infty} a_k$ converges as shown in (a).

c) $\langle (-1)^{k+1}/k \rangle$, $\langle 1 \rangle$

d) Let $a_{n_0} \ne 0$. Then $|a_{n_0} + 1| > q|a_{n_0}|$

$$|a_{n_0} + 2| > q^2|a_{n_0}| \quad \text{etc.}$$

Compare to a geometric series with $r = q > 1$.

e) There is a number u such that $1 < u < q$, and for $\varepsilon = q - u$, there is an n_0 such that for $n > n_0$, $|a_{n+1}/a_n| > u > 1$. By (d), a_k diverges.

h) If

$$\lim \left\langle \left| \frac{a_{n+1}}{a_n} \right| \right\rangle < 1$$

the series converges absolutely
If

$$\lim \left\langle \left| \frac{a_{n+1}}{a_n} \right| \right\rangle = 1$$

the series might converge conditionally or absolutely
If

$$\lim \left\langle \left| \frac{a_{n+1}}{a_n} \right| \right\rangle > 1$$

the series cannot converge, even conditionally

6. a) Geometric series (is $|r| < 1$?)
Comparison test for positive term series
Integral test for positive term series
Alternating series test
Ratio test for an arbitrary series

b) $1 + \dfrac{1}{10} - \dfrac{1}{2} + \dfrac{1}{3} + \dfrac{1}{30} - \dfrac{1}{4} + \dfrac{1}{5} + \dfrac{1}{50}$

$$+ \cdots + \frac{1}{2n+1} + \frac{1}{10(2n+1)} - \frac{1}{2n+2} + \cdots$$

Try to write others.

[3.6.1]

1. $\sum_{k=0}^{\infty} f_k(x)$ converges to $f_\lambda(x)$ on I, if and only if, for each $x \in I$ and each $\varepsilon > 0$, there is an integer $n_{x,\varepsilon}$ such that for all $n > n_{x,\varepsilon}$,

$$\left| f_\lambda(x) - \sum_{k=0}^{n} f_k(x) \right| < \varepsilon.$$

2. $\sum_{k=0}^{\infty} f_k(x)$ converges uniformly to $f_\lambda(x)$ on I, if and only if, for each $\varepsilon > 0$, there is an integer n_ε such that for all $x \in I$ and $n > n_\varepsilon$, $|f_\lambda(x) - \sum_{k=0}^{n} f_k(x)| < \varepsilon$.

3. a) i) $x = 2, 3$ b) i) $0, \frac{1}{2}$ d) i) $2, 0$
     ii) $x = 1$ ii) 1 ii) —
     iii) $[2, 3]$ or $[2, \infty)$ iii) $(-1, +1)$ iii) $(-\infty, \infty)$
     iv) $[2, 3]$ or $(-\infty, -2]$ iv) $(-1, +1)$ iv) $(-\infty, \infty)$
     v) $[2, 3]$ v) $[-\frac{1}{2}, +\frac{1}{2}]$ v) $[0, 4]$

4. c) $\lim |a_{n+1}(x - x_0)/a_n| < 1$ if $|x - x_0| < \lim |a_n/a_{n+1}|$

5. a) 1
     b) $1/1 - x = S_1(x)$, $S_2(x) = 1/2 - x$. [Note that carrying out the division $1/(1 - x)$ yields the series $\sum_{k=0}^{\infty} x^k = S_1(x)$.]
     d) Both diverge at both endpoints.
     e) $\rho = 1$

6. a) ρ remains the same.
     b) $S(x)$ converges for all x; $S'(x)$ does not converge at 0 or $n\pi$.

[3.7.1]

1. e) The series $\sum_{k=0}^{\infty} x^k/k^2$ is not defined because the term for $k = 0$ is not defined. Consider $\sum_{k=1}^{\infty} x^k/k^2$. The series representation of $f'(x)$ at $x = 1$ is the harmonic series; at $x = -1$ the series is $\sum_{k=1}^{\infty} (-1)^{k-1}/k$ and converges.
     f) Series (a) and (c) in 3 [3.6.1] do not have an interval domain of convergence.

2. Consider the partial sums.

[3.8.1]

1. b) $\rho = \infty$

5. $a_0 = e^{1/2}$, $a_1 = e^{1/2}$, $a_2 = \frac{1}{2}e^{1/2}$

6. $a_0 = 0$, $a_1 = 1$, $a_2 = 0$

8. d) Theorem 2.1.1

10. $a_0 = 0$, $a_n = (-1)^{n+1}/n$

[3.10.1]

2. d) $a_k = \dfrac{a_0}{k!}$, $a_0 = 1$

[3.10.2]

2. In the series expression for the differential equation the coefficients of each x^k must equal zero. Hence $0 \cdot a_0 = 0$, $0 \cdot a_1 = 0$ and $2 \cdot a_2 = 0$, and a_0 and a_1 are arbitrary constants, while $a_2 = 0$.

$$y = a_0 \left[1 + \frac{x^3}{3 \cdot 2} + \frac{x^6}{6 \cdot 5 \cdot 3 \cdot 2} \right.$$
$$\left. + \cdots + \frac{x^{3k}}{(3k)(3k-1)(3k-3)(3k-4) \ldots 3 \cdot 2} + \cdots \right]$$
$$+ a \left[x + \frac{x}{4 \cdot 3} + \frac{x}{7 \cdot 6 \cdot 4 \cdot 3} \right.$$
$$\left. + \cdots + \frac{x^{3k+1}}{(3k+1)(3k)(3k-2)(3k-3) \ldots 4 \cdot 3} + \cdots \right]$$
$$= a_0 S_1 + a_1 S2.$$

The Wronskian $W(a_0 S_1, a_1 S_2) \neq 0$ at $x = 0$: $W(a_0 S_1, a_1 S_2) \neq 0$ at $x = 0$.

3. a_0, a_1 arbitrary,

$$a_{k+2} = \frac{2(k-1)}{(k+1)(k+2)} a_k$$

[3.11.1]

5. $\bar{r}$ and $\bar{\theta}$ represent mean values which arise in the application of the mean-value theorem (see Section 2.22): $r_1 < \bar{r} < \bar{r}_2$ and $\theta_1 < \bar{\theta} < \theta_2$.

[3.12.1]

5. If

$$\frac{f_3''(t)}{f_3(t)} = \frac{T}{\sigma} \left(\frac{f_1''(r)}{f_1(r)} + \frac{f_1'(r)}{rf_1(r)} + \frac{1}{r^2} \frac{f_2''(\theta)}{f_2(\theta)} \right)$$

Then the expression on each side has a constant value c.

[3.13.1]

1.
3. } Check values with tables in Appendix C.

[3.15.1]

3. a)
$$r^2 \frac{f_1''(x)}{f_1(r)} + r \frac{f_1'(r)}{f_1(r)} - c \frac{\sigma}{t} r^2 = -\frac{f_2''(\theta)}{f_2(\theta)}$$

(see answer to 5 [3.12.1]). Then the expression on each side is a constant c_*. In this case c_* must be positive. Otherwise $f_2(\theta)$ would not be periodic, i.e., would not satisfy the condition $z(r, \theta, t) = z(r, \theta + 2\pi, t)$. Let $c_* = m^2$.

b) If $\cos m\theta = \cos m(\theta + 2\pi) = \cos(m\theta + 2m\pi)$, then m must be an integer.

6. $D_x J_0(x) = -J_1(x)$ and $D_x(x J_1(x)) = x J_0(x)$.

[3.15.2]

In the series solutions for the Bessel equations for $m = 2, 3$ and the general case m, the early coefficients a_k, $k < m$ are all zero. Note that in the first sentence in Section 3.16, $k < n$ should be $k < m$.

[3.18.1]

1. The last relation in the first sentence of Exercise 1 should be $\sigma = \sigma_0/r^2$.

   a) Use the relation $\sigma = \sigma_0/r^2$ in the two-dimensional wave equation (3.11.1) derived in Exercise 7 [3.11.1].
   b) $f''(r) + (1/r)f'(r) + (\omega^2 \sigma_0/T)(1/r^2)f(r) = 0$.
   d) $\omega = (2n + 1)(\pi/2)\sqrt{T/\sigma_0}\,(1/\ln r_0)$.
   e) $\omega_1 = 3\omega_0;\ \omega_2 = 5\omega_0$, etc.

[3.19.1]

4. a) In the solution $y = \sum_{k=0}^{\infty} a_k x^k$, a_0 and a_1 are both arbitrary, and the recurrence relation for $k = 2$ is
$$a_k = \frac{-2a_{k-2} + 3(k-1)a_{k-1}}{k(k-1)}.$$

   The two series $a_0(1 + \cdots)$ and $a_1(x + \cdots)$ are independent on some $I \supset \{0\}$ because the Wronskian at 0 is $a_0 a_1$.
   b) The series (with two arbitrary constants) or the two series mentioned in (a) are not easily recognizable.
   c) Initial conditions $y(0) = y_0;\ y'(0) = 0$ yield the series for $y = 2e^x - e^{2x}$ (obtained by algebraic methods, Theorem 2.12.1). Indeed, the general series with two arbitrary coefficients a_0, a_1 may be expressed as a linear combination $c_1 e^x + c_2 e^{2x}$, by imposing the initial conditions
$$y(0) = y_0 \Rightarrow a_0 = y_0, \qquad y'(0) = v_0 \Rightarrow a_1 = v_0.$$

   d) There are many possible bases, so that even a familiar function may not be easily recognizable when expressed in terms of certain series, which separately are not one of the familiar series.

[4.3.1]

3. a) The plane region determined by the points (2, 0, 0), (0, 2, 0), (0, 0, 2) and bounded by the vertical planes $x = \pm 2$, $y = \pm 2$
   b) The curve determined by the intersection of the plane region in (a) and the vertical plane $y(x) = x + 1$
   c) The curve determined by the intersection of the surfaces $y = 1 - x^2$ and $z = 2 - x - y$

[4.3.2]

Note a misprint: In the line above Exercises [4.3.2], the first f should have a bar over it: $\bar{f}$

1. b) $\left. \begin{array}{l} \bar{f}_1 : x \to f_1(x, x^3) = 3x^2 \\ \bar{f}_2 : x \to f_2(x, 0) = 0 \end{array} \right\}$ Both are continuous on R.

   c) $y_1 = 0 + \displaystyle\int_0^x 3t^2 \, dt = x^3$

   $y_2 = 0 + \displaystyle\int_0^x 0 \, dt = 0$

2. b) ω is not continuous on $[0, 3]$. g is continuous but not differentiable on $[0, 3]$.
   c) If $\hat{\omega}$ is continuous, then $\hat{g}$ is differentiable on $[0, 3]$.

3. a) $y'(x) = \bar{f}(x) = f(x, y(x))$: the continuity of $\bar{f} : x \to f(x, y(x))$ is a sufficient condition for the differentiability of the integral expression.

4. b) See Lemma 4.4.1.

[4.4.1]

2. *Note:* The set of differentiable solutions in $\mathscr{F}$ would give the same common set of solutions, but it is usual to state the weakest hypotheses that will support the arguments in a proof.

3. b) One such proof is included in Appendix D.

[4.5.1]

1. a) $s_0(x_0) = y_0$, $s_0'(x_0) = 0$
   $s_1(x_0) = y_0$, $s_1'(x_0) = f(x_0, y_0)$
   $y(x_0) = y_0$, $y'(x_0) = f(x_0, y_0)$
   b) At x_0, $s_1'(x)$ has the same slope as $y'(x)$. Therefore $s_1(x)$ should be closer to $y(x)$ than the horizontal function $s_0(x) = y_0$.

2. c) $y''(x_0) = s_2''(x_0)$

[4.6.1]

$y' = x - y, y(0) = 1$

1. a)

x:	0	0.4	0.8
y:	1	0.71	0.66

3. $s_0 = 1$; $s_1 = 1 - x + (x^2/2)$; $s_3 = 1 - x + x^2 - (x^3/3) + (x^4/4!)$
5. $s_4 = 1 - x + x^2 - (x^3/3) + (x^4/4 \cdot 3) - (x^5/5!)$
7. $y = x - 1 + 2e^{-x}$ (solution of a first-order linear equation, page 57–8)

[4.7.2]

Note a misprint: Equation (f) should be $y' = x - \ln y$.
1. (b) (c) (d); slope y' is not defined at (0, 1).
2. (b) y-axis (c) line $y = 1$ (d) y-axis (e) $x + y < 0$ (f) $y \leq 0$
4. a) (b) $s_0 = 0$, $s_1 = \ln x$, $s_2 = (\ln x - 1)(1 - x)$, for $x > 0$
5. (d) $s_0 = 0$, $s_1 = \frac{1}{2}(x^2 - 1)$, $s_2 = \frac{1}{4}(x^2 - 1) + \frac{1}{2}\ln x$, $x > 0$

(e) $s_1 = (\frac{2}{3})(x^{3/2} - 1)$, $x > 0$, $s_2 = \int_1^x (t + (\frac{2}{3})t^{3/2} - \frac{2}{3})^{1/2} \, dt$

b) (a) $s_1 = e^x - x$, $s_2 = 1 + (x^2/2)$, $s_3 = e^x - x - (x^3/3!)$
   (f) For the differential equation: $y' = x - \ln y$

$$s_1 = 1 + (x^2/2), \quad s_2 = 1 + \int_0^x t - \ln(1 + (t^2/2)) \, dt$$

c) (c) $s_1 = x + (x^2/2)$, $s_2 = \int_0^x t - (\frac{1}{2}(t + 1)^2 - \frac{3}{2})^{-1} \, dt$

s_2 is not defined when $(t + 1)^2 = 3$; $-1 - \sqrt{3} < x < -1 + \sqrt{3}$
7. (a), (b), (d)

[4.8.1]

1. If $\langle a_n \rangle = \langle 1/(n + 1) \rangle = \frac{1}{2}, \frac{1}{3}, \ldots$, then $\lim \langle 1/(n + 1) \rangle = 0$, and the limit 0 is not in the open interval (0, 1).
   If $\langle a_n \rangle = \langle (\frac{1}{2}) + (1/(n + 2)) \rangle$, then the limit, $\frac{1}{2}$, is in the open interval (0, 1).
   If $b_n = \langle 1/n \rangle$, then the limit 0 is in the closed interval [0, 1].

   Hint: Consider any number $r \notin [0, 1]$. Show that r is not a limit point of $\langle a_n \rangle$ if every $a_n \in [0, 1]$; that is, find an $\varepsilon > 0$, such that $|r - a_n| > \varepsilon$ for all n.

2. a) *Hint:* The function need not be continuous. Consider

$$f: x \to 1/x \quad \text{for} \quad 0 < x \le 1$$
$$\to 0 \quad \text{for} \quad x = 0$$

  b) $f: x \to 1/x$ on $(0, 1)$

  c) A continuous function on a closed interval $[a, b]$, where $a, b \in R$ is bounded. But note that the interval $(-\infty, \infty)$ is considered both open and closed; and continuous, unbounded functions such as $y = e^x$ do exist on this closed, unbounded, interval.

3. f continuous on $[a, b]$ indicates that f is bounded on $[a, b]$, that is, there is a positive number M such that $|f(t)| \le M$ for $t \in [a, b]$. Then $\int_a^b f(t)\,dt \le \int_a^b M\,dt = M(b - a)$.

4. a) $f(x, y) = 1/(y - y_0)$

  b) Impossible

[4.10.1]

3. $|s_k(x) - s_{k-1}(x)| \le ML^{k-1}\alpha^k$, for $x \in I_\alpha$

4. The geometric series with ratio $L\alpha$ converges if $|L\alpha| < 1$; that is, if $\alpha < 1/L$.

5. Each s_n is bounded by

$$s_0 + M\alpha \sum_{k=1}^{\infty} (\alpha L)^{k-1}$$

and for $\alpha < 1/L$, the series has a finite sum.

6.
$$|s_\lambda - s_n(x)| = \left| \sum_{k=n+1}^{\infty} (s_k - s_{k-1}) \right| \le M\alpha \sum_{k=n+1}^{\infty} (L\alpha)^{k-1}$$

and the sum of this geometric series is easily determined for $\alpha < 1/L$.

7. If f is continuous and satisfies a Lipschitz condition on $P(A, B)$, then the Picard approximation functions are well defined for $x < \alpha = \min \{A, B/M\}$, and converge to a solution function on the interval I_α, where $\alpha < 1/L$.

8. $|f(t, y_2) - f(t, y_1)| = |t - y_2 - (t - y_1)| = |y_1 - y_2| \le 1|y_2 - y_1|, \quad L = 1.$

  I_α, in Exercise 4 above, is restricted to $\alpha < 1/L = 1$. Yet the solution $y = x - 1 + 2e^{-x}$ in Exercise 7 [4.6.1] is valid on R. Perhaps the choice of bounds was too careless.

[4.10.2]

1. a) $|s_2(x) - s_1(x)| = \left| \int_{x_0}^{x} f(t, s_1(t)) - f(t, s_0)\,dt \right| \le L \left| \int_{x_0}^{x} |s_1(t) - s_0|\,dt \right|$

$$\le LM \left| \int_{x_0}^{x} |t - x_0|\,dt \right|$$

3. $(M/L)L^k|x - x_0|^k/k!$

5. a) $f(x) = (M/L)e^{L|x-x_0|}$

   b) $x \in R$

   c) $|x - x_0| < \alpha$

[4.11.1]

1. a) If $\{(x, s_{k-1}(x)): x \in I_\alpha\}$ is in $P(A, B)$, then $f(x, s_{k-1}(x))$ is well defined, continuous, and bounded by M. Therefore

$$|s_k(x) - y_0| = \left| \int_{x_0}^x f(t, s_{k-1}(t)) \, dt \right| \le M|x - x_0|.$$

Use an induction argument; the relation has been established for $|s_1(x) - x_0|$. (See page 186.)

[4.12.1]

3. a) $f(x, y(x)) = 3xy^{1/3}$ does not satisfy a Lipschitz condition on P_2.

   b) The inequality in this exercise should read

$$|3xy_2^{1/3} - 3xy_1^{1/3}| \le L|y_2 - y_1|.$$

   Suppose $L = 100$. Consider points $(x, y_1) = (\frac{1}{3}, 0)$ and $(x, y_2) = (\frac{1}{3}, 0.008)$. Then $|0.2 - 0| \le 0.008$. In general, for $y_1 = 0$, choose y_2 such that $y_2^{-2/3} > L/3|x|$, $x \ne 0$.

   c) Consider the region $P = \{(x, y): |x| < 1, 1 < y < 2\}$

[4.14.1]

2. a) Any region

   b) $\{(x, y_1, y_2): x \in I, y_1, y_2 \text{ unrestricted}\}$

   c) Region in (b)

3. b) $\vec{f}_i: x \xrightarrow{(i, y_1, y_2)} (x, y_1, y_2) \xrightarrow{f_i} f_i(x, y_1, y_2)$

6. $s_{1,0} = y_{1,0}, \qquad s_{2,0} = y_{2,0}$

$$s_{1,1} = y_{1,0} + \int_{x_0}^x f_1(t, s_{1,0}, s_{2,0}) \, dt = y_{1,0} + \int_{x_0}^x y_{2,0} \, dt$$

$$s_{2,1} = y_{2,0} + \int_{x_0}^x -p(t)y_{2,0} - q(t)y_{1,0} + r(t) \, dt$$

$$s_{1,2} = y_{1,0} + \int_{x_0}^x s_{2,1} \, dt$$

$$s_{2,2} = y_{2,0} + \int_{x_0}^x -p(t)s_{2,1} - q(t)s_{1,1} + r(t) \, dt$$

[4.14.3]

1. a) $s_{1,0} = y_{1,0};$ $s_{1,1} = y_{1,0} + \int_{x_0}^{x} f_1(t, y_{1,0}, y_{2,0})\, dt$

$s_{1,1} - s_{1,0} = \int_{x_0}^{x} f_1(t, y_{1,0}, y_{2,0})\, dt$

   b) The f_i are continuous and thus bounded on the closed and bounded region, P. If M_1 is a bound for both f_1 and f_2, let $M = 2M_1$.

2. a) $|s_{1,2} - s_{1,1}| = \left| \int_{x_0}^{x} f_1(t, s_{1,1}, s_{2,1}) - f_1(t, s_{1,0}, s_{2,0})\, dt \right|$

$\le L_1 \left| \int_{x_0}^{x} |s_{1,1} - s_{1,0}| + |s_{2,1} - s_{2,0}|\, dt \right|$

$\le L_1 \left| \int_{x_0}^{x} M|t - x_0|\, dt \right| \le L_1 M |x - x_0|^2 / 2.$

$|s_{2,2} - s_{2,1}| \le L_2 M |x - x_0|^2 / 2$

4. c) Compare the series with the exponential series for $(M/L_1 + L_2)e^{(L_1 + L_2)|x - x_0|}$.

5. a) Both sequences converge uniformly on I.
   b) Each $s_{i,\lambda}$ is defined and continuous on I.

[4.14.4]

3. Vector notation would simplify the writing of such a proof for order n.

[4.15.1]

4. The Euler method, in the notation of page 10, is
$$y(x + h) = y(x) + hy'(x).$$
In the notation of page 199, $x = x_0$, $x + h = x_1$, and $h = x_1 - x_0$. Thus $y(x_1) = y(x_0) + y'(x_0)(x_1 - x_0)$, and these are the first two terms of a Taylor series for $y(x_1)$ at x_0.

5. The modified Euler method, in the notation of page 10, is
$$y(x + h) = y(x) + (h/2)(y'(x) + y'(x + h)).$$
And the Euler method in Exercise 4, applied to $y'(x + h)$, is
$$y'(x + h) = y'(x) + hy''(x).$$
Substituting this last expression in the modified method,
$$y(x + h) = y(x) + hy'(x) + (h^2/2)y''(x).$$
In the notation of page 199 (see answer to Exercise 4 above), the last expression becomes
$$y(x_1) = y(x_0) + y'(x_0)(x_1 - x_0) + \frac{y''(x_0)}{2!}(x_1 - x_0)^2,$$
and these are the first three terms of the Taylor series for $y(x_1)$ at x_0.

LIST OF SYMBOLS

$\in$, is an element of, 1

$\{\quad\}$, set, 1

$|\quad|$, absolute value, 3

$\cap$, intersection, 26

$\supset$, contains, 42

$\circ$, composite, 43

$\begin{vmatrix} a & b \\ c & d \end{vmatrix}$, determinant, 67

$\propto$, proportional, 84

$\langle\quad\rangle$, sequence, or ordered set, 122

$\Rightarrow$ implies, 178, 210

$\doteq$, approximately equal to, 10

$\blacksquare$, end of proof, 26

$\star$ Indicates exercises, sections, or discussions which are difficult or theoretical and may be omitted without loss of continuity, 11

a, acceleration, 12

A, frontal area, 22

amount of C^{14}, 39

amplitude, 73

$A(x) = \int_a^x f(u)\,du$, integral function, 7

$\alpha(t)$, radian measure, 75

α_i, zero of J_0, 150

α_n^m, nth positive zero of J_m, 157

β, phase displacement, 73

C, capacitance, 105

C^{14}, carbon-14, 38

c_i, roots of indicial equation, Exercise 1(c) [3.16.1], 155

$\mathscr{C}$, set of continuous functions whose graphs are in P, 177

D, differential operator, 62, 63

$D_x y = dy/dx = y'(x)$, derivatives, 4

$\partial y/\partial x = y_x,\ \partial^2 y(x, t)/\partial x^2$, partial derivatives, 110

∇^2, Laplacian operator, 147

E, error, 12, 138

$\overrightarrow{EP}$, ray, 47

e, $\ln e = 1$, ($e \doteq 2.71$), 3

  eccentricity, 51

$\exp u = e^u$, 58

$f: x \to f(x)$, function, 1

$f[I]$, range $\{f(x): x \in I\}$, 1

0_f, r_f, constant functions, 2

$f^{\leftarrow}$, inverse function, 2

$f^{(n)} = y^{(n)}$, nth derivative, 12

$f^{[n]} = y^{[n]}$, nth numerical estimate of $f(x) = y(x)$, 199

$f_\lambda = \lim \langle f_n \rangle$, 125

F, force, 16

F_g, gravitational force, 17

F_r, resisting force, 18

$\mathscr{F} = \{w \text{ on } I_w: \{(x, w(x)): x \in I_w\} \subset P\}$, 174

g, constant acceleration, 13

$|g|$, gravitational force per unit mass, 17

G, constant in the gravitational law, 116

glb, greatest lower bound of set $\{\int_{y_0}^{y} 1/q(u)\, du + x_0: y \in J\}$, 26

Γ, gamma function, 171

γ, Euler's constant, 125

h, height, 12

H, constant of proportionality in Hooke's law, 84

$i: x \to x$, identity function, 2

I, interval, 1

I_H, interval domain for Hooke's law, 85

I_α, common domain for Picard functions, 185–186

J_0, Bessel function of order 0, 149

J_1, Exercise 5 [3.15.1], 153

J_m, 155

k, drag coefficient, 18

L, inductance, 105

  limit, 123

  linear operator, 63

  Lipschitz number, 187

L_c, linear operator with constant coefficients, 86

lub, least upper bound of set $\{\int_{y_0}^{y} 1/q(u)\, du + x_0: y \in J\}$, 26

$\lim_{t \to +0}$, limit as t (> 0) approaches 0, 14

λ, wavelength of light, 163

m, mass, 17

  root of auxiliary algebraic equation, 86–87

$\max_{x \in I_\alpha} |s_\lambda(x) - s_\beta(x)|$, maximum value in the set $\{|s_\lambda(x) - s_\beta(x)|: x \in I_\alpha\}$, 192 min $\{A, B/M\}$, minimum value of A and B/M, 194

μ, micron, 10^{-6} cm, 163

$\mathcal{O}$, operator, 62

ω, angular frequency, 73

ω_0, natural frequency of oscillator, 97

$\omega_{m,n}$, frequency of nth normal mode with m nodal circles (characteristic value), 157

$P(\omega)$, magnification factor, 98

$\mathbf{P} = (x, y)$, vector, 50, 116

$P(A, B) = \{(x, y): |x - x_0| \leq A, |y - y_0| \leq B\}$, rectangular region, Exercise 4 [4.8.1], 185

q, charge, 105

$\dot{q} = $ current (I), 105

R, resistance of wire, 105

    set of real numbers, 1

R^+, positive reals, 1

$R_n(x)$, remainder, 139

ρ, atmospheric density, 22

ρ, radius of convergence, Exercise 4 [3.6.1], 133

$S_n = \sum_{k=1}^n a_k$, partial sum, 126

s_p, particular solution of $L_c(y) = r(x)$, 95

$\dot{s}(t)$, derivative with respect to t, 2

S_n^+, see Exercise 2 [3.5.2], 131, and answer

S_n^-, see Exercise 2 [3.5.2], 131, and answer

$s_n(x) = y_0 + \int_{x_0}^x f(t, s_{n-1}(t))\, dt$, 173

$s_\lambda = \lim \langle s_n \rangle$, 188

$\sum_{k=0}^{\infty} a_k x^k$, power series, 88, 120

$\sum_{k=1}^{\infty} a_k$, numerical series, 126

$\sum_{k=1}^{n} c_k f_k$, linear combination of functions, 64

V, voltage, 105

v, velocity, 12

$v_f: t \to v$, velocity as a function of time, 47

$\hat{v}: r \to v$, velocity as a function of position, 47

$W(f_1, f_2)$, Wronskian determinant, 67

$\psi_{m,n}(r, \theta)$, characteristic function, 157

$\bar{x}$, mean value $(x_i < \bar{x}_i < x_{i+1})$, 111

Y_0, Bessel function of the second kind, 168–169

BIBLIOGRAPHY

Abell, George, *Exploration of the Universe,* second edition, pages 530–536. New York: Holt, Rinehart, and Winston, 1969

Andrade, E. N. da C., *Sir Isaac Newton,* Garden City, N.Y.: Doubleday, 1958

Astin, Allen V., "Standards of Measurement," *Sci. Amer.* **218,** 6, 50–62, June 1968

Ball, Sir R. S., *Time and Tide,* London: Society for Promoting Christian Knowledge, 1889

Bartrum, C. O., "Time: Its Determination, Measurement, and Distribution," *Splendour of the Heavens,* edited by T. E. R. Phillips and W. H. Steavenson. London: Hutchinson, 1923

Bell, E. T., *Men of Mathematics,* New York: Simon & Schuster, 1937

Bergmann, L., "Experiments with Vibrating Soap Membranes," *J. Acoust. Soc. Am.* **28,** 1043f, November 1956

Blanchard, D. C., *From Raindrops to Volcanoes,* Anchor S50. Garden City, N.Y.: Doubleday, 1967

Bowman, Frank, *Introduction to Bessel Functions,* New York: Dover, 1958

Breasted, J. H., "The Beginnings of Time Measurement," *Time and Its Mysteries,* Series I. New York: New York University Press, 1936

British Astronomical Association Handbook, London, 1971

Brown, L. A., "The Longitude," *The World of Mathematics,* Vol. II, pages 780–819, New York: Simon & Schuster, 1956

Chladni, Ernst, *Die Akustik,* 1802

Clancy, E. P., *The Tides,* Science Studies Series, Anchor S56. Garden City, N.Y.: Doubleday, 1968

Coddington, Earl A., *An Introduction to Ordinary Differential Equations,* Englewood Cliffs, N. J.: Prentice-Hall, 1961

Cohen, H., and G. Handelman, "On the Vibration of a Circular Membrane with Added Mass," *J. Acoust. Soc. Am.* **29,** 2, 1957

Conte, S. D., *Elementary Numerical Analysis,* New York: McGraw-Hill, 1965

Cooper, P. W., "Through the Earth in Forty Minutes," *Amer. J. Phys.* **34,** 1, pages 68–70, 1966. The letters concerning this article, in Volume 34:8, pages 701–704, make interesting reading.

Courant, R., *Differential and Integral Calculus,* Vol. 1, second edition, New York: Interscience, 1937

Deevey, E. S., Jr., "Radioactive Dating," *Sci. Amer.* **186,** 2, 24–28, February 1952

Edwards, L. K., "High-Speed Tube Transportation," *Sci. Amer.* **213,** 2, 30–40, August 1968

Ferguson, C. W., "Bristlecone Pine; Science and Esthetics," *Science* **159**, 839–846, February 23, 1968

Feynman, R. P., R. B. Leighton, and M. Sands, *The Feynman Lectures on Physics,* Vol. 1, Reading, Mass.: Addison-Wesley, 1965

Fireman, Edward L., "The Lost City Meteorite," *Sky and Telescope* **39**, 3, 158, March 1970

Galilei, Galileo, *Dialogues Concerning Two New Sciences,* "First Day," "Third Day," translated by H. Crew and A. DeSalvio. New York: Macmillan, 1914

Gardner, M., letter, *Sci. Amer.* **213**, 3, 10, September 1965

Ghosh, R. N., "Note on Indian Drums," *Phys. Rev.* **20**, 526, 1922

Greenspan, Donald, *Theory and Solution of Ordinary Differential Equations,* New York: Macmillan, 1960

Hawkes, Jacquetta, "New Dates for Old Times," *The Sunday Times,* London, page 8, December 14, 1969

Heide, Fritz, *Meteorites,* Phoenix Science Series 522. Chicago: University of Chicago Press, second edition, 1965. Translation of *Kleine Meteoritenkunde,* second edition, Berlin: Springer Verlag, 1957

Hochstadt, Harry, *Differential Equations,* New York: Holt, Rinehart, and Winston, 1964

Holton, G., *Introduction to Concepts and Theories in Physical Science,* Reading, Mass.: Addison-Wesley, 1952

Hoyle, F., *The Black Cloud,* P37, Perennial Library. New York: Harper and Row, 1957

Humphreys, W. J., *Physics of the Air,* third edition. New York: McGraw-Hill, 1940

Hurley, P. M., "Radioactivity and Time," *Sci. Amer.* **181**, 2, 48–51, August 1949

Jahnke, Eugene, Fritz Emde, and Friedrich Losch, *Tables of Higher Functions,* sixth edition, pages 132–134, 158–163, 192–195. New York: McGraw-Hill, 1960

Josephs, Jess J., *Physics of Musical Sound,* Momentum Book 13, page 14. Princeton, N.J.: Van Nostrand, 1967

Joshua, **X:** 11, Old Testament, *King James Version of the Bible*

Keisch, Bernard, "Dating Works of Art Through Their Natural Radioactivity: Improvements and Applications," *Science* **160**, pages 413–415, April 26, 1968

Kelly, L. G., *Handbook of Numerical Methods and Applications,* Reading, Mass.: Addison-Wesley, 1967

Koestler, Arthur, *The Sleepwalkers,* London: Hutchinson, 1959

Libby, Willard F., *Radiocarbon Dating,* second edition. Chicago: University of Chicago Press, 1955

Libby, W. F., E. C. Anderson, and J. R. Arnold, "Age Determination by Radiocarbon Content: Worldwide Assay of Natural Radiocarbon," *Science* **109**, pages 227–228, March 4, 1949

Lyons, Harold, "Atomic Clocks," *Sci. Amer.* **196**, 2, 71–82, February 1957

Manchester Chronicle, April 16, 1831 (collapse of Broughton Suspension Bridge)

Manchester Guardian, April 16, 1831 (collapse of Broughton Suspension Bridge)

McCrosky, Richard E., "The Lost City Meteorite Fall," *Sky and Telescope* **39**, 3, 154–158, March 1970

McDonald, J. E., "The Shape of Raindrops," *Sci. Amer.* **190**, 2, 64, February 1954

Melvin, M. A., and S. Edwards, Jr., "Group Theory of Symmetric Molecules, Membranes, and Plates," *J. Acoust. Soc. Am.* **28**, 201, 1956

Milham, W. I., *Time and Timekeepers*. New York: Macmillan, 1923

Millikan, R. A., "Time," *Time and Its Mysteries,* Series I. New York: New York University Press, 1936

Monahan, M. A., and K. Bromley, "Vibration Analysis by Holographic Interferometry," *J. Acoust. Soc. Am.* **44,** 5, 1225–31, 1968

Morse, P. M., *Vibration and Sound*. New York: McGraw-Hill, 1936

Muller, P. M., and W. L. Sjogren, "Mascons: Lunar Mass Concentrations," *Science* **161,** 3842, 680–684, and cover, August 16, 1968

Munkres, J. R., *Elementary Linear Algebra,* Reading, Mass.: Addison-Wesley, 1964

New York Times, May 22, 1854 (collapse of Wheeling, Ohio, suspension bridge)

New York Times, page 1 of Travel Section, May 22, 1960

New York Times Index, "Airlines, U.S., accidents," 1959, 1960, 1961

Newton, Sir Isaac, *Mathematical Principles,* Definition V, Cajori's revision of Motte's English Translation (1729), Berkeley, Calif.: University of California Press, 1934

Noble, J. V., and de Solla Price, Derek J., "The Water Clock in the Tower of the Winds," *American Journal of Archaeology* **72,** 345–355, plates 111–118, 1968

Park, David, *Contemporary Physics,* Chapter 1. New York: Harcourt, Brace, and World, 1964

Petrie, W. M. F., "Review of Daressy, G., 'Deux Clepsydras Antiques' " (*Bull. Inst. Egypt* V, IX, 1915), *Ancient Egypt,* pages 42–44, 1917

Physical Science Study Committee, *Physics,* first edition. Boston: D. C. Heath, 1960

Pontryagin, L. S., *Ordinary Differential Equations*. Reading, Mass.: Addison-Wesley, 1962

Press, Frank, "Resonant Vibrations of the Earth," *Sci. Amer.* **213,** 5, 28–37, November 1965

Putnam, J. L., *Isotopes,* Harmondsworth, England: Penguin Books, 1960

Rainville, Earl D., *Intermediate Differential Equations*. New York: Wiley, 1943

Rainville, Earl D., *Special Functions*. New York: Macmillan, 1960

Ramakrishna, B. S., "Modes of Vibration of the Indian Drum Dugga or Lefthand Thabala," *J. Acoust. Soc. Am.* **29,** 2, 1957

Ramakrishna, B. S., and M. M. Sondhi, "Vibrations of Indian Musical Drums Regarded as Composite Membranes," *J. Acoust. Soc. Am.* **26,** 523–529, 1954

Renfrew, Colin, "Carbon 14 and the Prehistory of Europe," *Sci. Amer.* **225,** 4, 63–72, October 1971

Ritchie-Calder, Lord, "Conversion to the Metric System," *Sci. Amer.* **223,** 1, 17–25, July 1970

Robinson, N. W., and R. W. B. Stephens, "On the Behavior of Liquid Films in a Vibrating Air Column," *Phil. Mag.* **17,** pages 27–33, 1934

Rogers, Eric M., *Physics for the Inquiring Mind*. Princeton, N. J.: Princeton University Press, 1960

Romans, Bernard, *Concise Natural History of East and West Florida,* New York, 1775

Runcorn, S. K., "Corals as Paleontological Clocks," *Sci. Amer.* **215,** 4, 26–33, October 1966

Russell, John Scott, "On the Vibration of Suspension Bridges and Other Structures, and the Means of Preventing Injury from This Cause," *Transactions of the Royal Scottish Society of Arts,* Vol. I, 1841

Scarborough, J. B., *Differential Equations and Applications,* Baltimore: Waverly Press, 1965

Scarborough, J. B., *Numerical Mathematical Analysis,* sixth edition.   Baltimore, Johns Hopkins Press, 1966

Shapiro, A. H., *Shape and Flow,* Anchor S21.   Garden City, N.Y.: Doubleday, 1961

Simpson, Colin, "How the Golden Armada Went Down," *Sunday Times Weekly Review,* page 41, London, December 11, 1966

Simpson, Colin, "Real Eight, Inc.," *The Sunday Times Magazine,* pages 8–13, London, December 18, 1966

Steinman, David, and Sara Ruth Watson, *Bridges and Their Builders.* New York: Dover, 1957

Symposium on Radioactive Dating, Athens, 1962. Vienna: International Atomic Energy Agency, 1963

"Tacoma Narrows Bridge Collapse," film loop 80–218, Ealing Film Loops, Cambridge, Mass., 1940

Tyndall, John, *Sound,* third edition.   New York: D. Appleton, 1885

Waller, Mary D., *Chladni Figures.* London: G. Bell, 1961

Waller, Mary D., *Chladni Plates.* London: Staples Press, 1960

Watson, G. N., *A Treatise on the Theory of Bessel Functions,* second edition.   New York: Macmillan, 1944

Willis, J. T., K. A. Browning, and D. Atlas, "Radar Observations of Ice Spheres in Free Fall," *J. Atmos. Sci.* **21,** pages 103–108, 1964

Wolfe, James, "A Proof of Taylor's Formula," *Am. Math. Month.* **60,** page 415, 1953

INDEX

INDEX

Theorem 2.12.1 [on solutions of $L_c(y) = y'' + py' + q(y) = 0$ where p and q are constant]

The general solution of the equation $L_c(y) = 0$ may be written in the form

$$f(x) = e^{ax}(c_1 \cos bx + c_2 \sin bx)$$

if the auxiliary equation has roots $a \pm bi$, $b \neq 0$;

$$f(x) = e^{ax}(c_1 + c_2 x)$$

if the auxiliary roots are a, a;

$$f(x) = c_1 e^{ax} + c_2 e^{bx}$$

if the auxiliary roots are $a \neq b$.

p. 90

Theorem 2.17.1 [on solutions of $L_c(y) = y'' + py' + qy = r(x)$ where p and q are constant]

Let r be continuous on the interval I.
Let s_p be a particular solution of the equation $L_c(y) = r(x)$ on I.
Let w be any other solution of $L_c(y) = r(x)$ on I.
Let $f(x) = c_1 y_1(x) + c_2 y_2(x)$ be the general solution of the corresponding homogeneous equation $L_c(y) = 0$.

p. 95

Then

$$w(x) = c_1 y_1(x) + c_2 y_2(x) + s_p(x) \quad \text{on} \quad I \quad \text{for some} \quad c_1, c_2 \in R.$$

The general solution is

$$y(x) = c_1 y_1(x) + c_2 y_2(x) + s_p(x)$$

[on particular solutions s_p of $L_c(y) = r(x)$, p, q constant]
see *method of undetermined coefficients*, section 2.17

pp. 96–97

[on solutions of $L(y) = 0$, p, q nonconstant]
see series solutions, sections 3.10, 3.12, 3.13, 3.15, 3.16, 3.19

pp. 141–143

148–149
153–156
166–170